Veterinary Virology

Veterinary Virology

Tom Richards

Larsen & Keller
www.larsen-keller.com

Veterinary Virology
Tom Richards
ISBN: 978-1-64172-700-6 (Hardback)

Larsen & Keller
Published by Larsen and Keller Education,
5 Penn Plaza,
19th Floor,
New York, NY 10001, USA

Cataloging-in-Publication Data

Veterinary virology / Tom Richards.
p. cm.
Includes bibliographical references and index.
ISBN 978-1-64172-700-6
1. Veterinary virology. 2. Agricultural virology. 3. Veterinary microbiology.
4. Virus diseases. I. Richards, Tom.
SF780.4 .V48 2022
636.089 601 94--dc23

For more information regarding Larsen and Keller Education and its products, please visit the publisher's website www.larsen-keller.com

Table of Contents

Preface

Veterinary virology is an important branch within veterinary medicine which studies viruses in non-human animals. There are various types of viruses, which are studied within this discipline such as rhabdoviruses, pestiviruses and coronaviruses. Rhabdoviruses are a family of viruses, which have a single strand negative sense RNA. Some of the pathogens within this family are potato yellow dwarf virus, rabies virus and vesicular stomatitis virus. Influenza is a disease which is caused by an influenza virus. The two prominent types of influenza are avian and swine influenza. This book provides significant information of this discipline to help develop a good understanding of veterinary virology and related fields. It consists of contributions made by international experts. Coherent flow of topics, student-friendly language and extensive use of examples make this book an invaluable source of knowledge.

A foreword of all Chapters of the book is provided below:

Chapter 1 - Veterinary virology is the branch of veterinary medicine which deals with study of impacts of viruses on animals. Rhabdoviruses, foot-and-mouth disease virus, pestiviruses, arteriviruses, coronaviruses, bluetongue virus, culturing viruses, etc. are some of the viruses studies in this field. This is an introductory chapter which will briefly introduce all the significant aspects of veterinary virology; **Chapter 2** - Zoonosis is a type of viral disease that spreads from animals to humans. It includes buffalo pox virus, yaba monkey tumor virus, mousepox, rabbitpox, swinpox virus, fibroma virus, pseudocowpox virus, asafarviradae, iridoviridae, etc. This chapter has been carefully written to provide an easy understanding of zoonotic viruses and diseases; **Chapter 3** - Herpesvirales is a dsDNA virus which is recognized by icosahedral capsid enclosed in a lipid envelope. Some of its types include bovine herpesvirus, equine herpesvirus type 1, felid herpesvirus, suid herpesvirus, duck virus enteritis, oncorhynchus masou virus, etc. These diverse viruses and diseases in animals related to herpesvirales virus have been thoroughly discussed in this chapter; **Chapter 4** - A type of virus, of size ranging between 90-100nm, having no outer envelope with double stranded DNA genome is referred to as adenoviridae virus. Canine Adenovirus, Hydropericardium Syndrome, Egg Drop Syndrome Virus, etc. are some of the viruses that fall in its domain. The topics elaborated in this chapter will help in gaining a better perspective about adenoviridae viruses; **Chapter 5** - Papillomaviridae and Polyomaviridae families have distinct biological features, but have similarities in virion structure, mechanism of replication and cell cycle regulation. Canine papillomavirus, papillomavirus of birds, polyomavirus, bovine polyovirus infection, etc. are some of its examples. All these diverse papillomaviridae and polymaviridae viruses have been carefully analyzed in this chapter; **Chapter 6** - Parvoviridae virus comprises of sixty copies of a single protein sequence, called VP1, VP2 etc. Circoviridae virus is non-enveloped consisting icosahedral and round geometries.

This chapter delves into the study of feline panleukopenia virus, canine parvovirus, goose parvovirus, porcine circovirus, chicken anemia virus, etc. to provide in-depth knowledge of the subject.

I would like to thank the entire editorial team who made sincere efforts for this book and my family who supported me in my efforts of working on this book. I take this opportunity to thank all those who have been a guiding force throughout my life.

Tom Richards

1

Introduction to Veterinary Virology

Veterinary virology is the branch of veterinary medicine which deals with study of impacts of viruses on animals. Rhabdoviruses, foot-and-mouth disease virus, pestiviruses, arteriviruses, coronaviruses, bluetongue virus, culturing viruses, etc. are some of the viruses studies in this field. This is an introductory chapter which will briefly introduce all the significant aspects of veterinary virology.

Veterinary virology and prion research has contributed greatly to our understanding of viruses and prions, the infections and diseases that they cause and their epidemiology and ecology. The importance of veterinary virology and prion research to animal health and the quality of our food supply is obvious and widely appreciated. However, the contributions of veterinary virology and prion research are not limited to animal health and the quality of our food supply. Veterinary virology and prion research has also had several major direct impacts in our understanding of human diseases. For example, many important human viruses or prions (or agents very closely related to them) were discovered by veterinarians, veterinary virologists, or prion researchers. Most human viruses and prions originate from animal agents that cross the so-called "species barrier" and became infectious to human beings. Research in animal viruses is therefore directly related to emerging human diseases. Veterinary virology and prion research has also led to many developments that directly impact human health.

For example, both traditional and modern rational vaccines were developed, or heavily influenced, by veterinary research. Veterinary viruses have also been used as surrogates of human viruses in the discovery and development of novel antiviral drugs and vaccines against human pathogens. Moreover, pathogenesis of human viral infections is also often studied by analyzing that of closely related veterinary viruses.

The importance of veterinary viruses and veterinary virology to human health and well-being is great. We will also include the prions, which together with viruses are the smallest transmissible infectious agents and were consequently studied together with viruses until very recently. Veterinary virology is of course of paramount importance to animal health, which is in turn essential for the safety of our food supply. The viruses that infect any of the animal species used as major sources of protein in the human diet can all threaten human health and wellbeing. There is a low but concrete probability of viral pandemics in any of these animal species, which would have drastic impacts on the supply of proteins for the human diet. The latest outbreak of food and mouth disease virus (FMDV) in the UK in 2001 is a good example of a geographically restricted but major disruption to the production of animal protein. In only 10 months, 581,802 heads of cattle (approximately 1 in 20), 3,487,014 of sheep (approximately 1 in 10), and 146,145 pigs in the UK were lost to the outbreak. Veterinary virology is therefore directly pertinent to the security of the food supply, importance which is obvious and widely recognized. However, veterinary virology also has direct impacts in human health, impacts which are not so often equally well appreciated.

Viruses and prions pose a major direct threat to global human health. Many viruses affecting humans are actually zoonotic, that is they infect animals and are maintained in animal reservoirs from which they spread to human beings. Old human viruses such as influenza are continuously re-entering into the human population from their animal reservoirs, with a low but concrete probability of producing another global pandemic. Such threat was just highlighted by the 2009 H1N1 strain, which most likely was recombined in pigs before being transmitted to humans. Other more recently identified viruses, such as hepatitis E (HEV), are most likely periodically transmitted to humans directly from their animal reservoirs. New human viruses or prions such as SARS, Nipah, or bovine spongiform encephalopathy (BSE, also known as "mad cow disease") also periodically enter into the human population from animal reservoirs. It is currently accepted in Public Health that "most new human infections are of animal origin." Before they enter into human populations, most viruses are therefore perpetuated in susceptible animals. Major outbreaks of animal disease can also provide a large source for the introduction of novel viral pathogens in the human population. For example, the large outbreak of BSE eventually resulted in the introduction of a new prion disease to humans, variant Creutzfeldt-Jakob disease (vCJD). Veterinary virology is consequently directly pertinent to emerging and established infectious human diseases.

Several animal viruses are also closely related to human viruses, being transmitted by similar routes, producing similar diseases, and being controlled by similar immune responses. Such animal viruses are therefore excellent models to study pathogenesis of, and therapeutic and preventive approaches against, the equivalent human viruses. This approach has been especially useful in the development of novel vaccines. It has also been useful to understand the pathogenesis of, and even developing drugs against, human viruses that infect no practical animal models, such as HCV or HIV. Veterinary virology is therefore also directly pertinent to our understanding of human disease, and to the development and testing of novel antiviral therapeutics or vaccines.

Veterinary Sciences in the Foundations of Virology and in Ongoing Virus Discovery

Viral diseases constitute a continuous threat to human health. Viruses such as influenza continue to produce periodic epidemics and kill scores of human beings. Influenza is estimated to kill more than 36,000 people per year in the USA and 500,000 in the world. The flu pandemic of 1918 is estimated to have killed up to 1 in 20 people living at the time. The recent emergency of a novel H1N1 strain proves that new pandemics are likely to occur in the not so distant future. Viral diarrhoeas are one of the major causes of infant death in large parts of the world. They are estimated to cause 5 to 20 million infant deaths in Africa and Latin America per year. HIV has already killed more than 25 million people and it currently infects more than 33 million more (23 million of them in Sub-Saharan Africa). Hepatitis C virus (HCV) infects 3 to 4 million more people every year, resulting in 170 million infected people in the world and an estimated 1.7 million cases of liver cancer. New viral diseases are constantly threatening to enter into the human population and produce serious pandemics. Virology research is therefore essential to human health.

Veterinary sciences have played major roles in virology research since the very origins of the discipline. The search for novel infectious agents started immediately after the seminal works by Pasteur, Koch and their contemporaries had established the bases for the microbial theory of infectious diseases. A purely technical development, the Chamberland-Pasteur unglazed porcelain (ultra) filter, resulted in the discovery that some infectious agents were much too small to be regular bacteria. Bacteria, parasites or fungi could be identified under the optic microscope, but infectivity was then the only property that allowed the identification and characterization other infectious agents. The infectivity of infectious agents of animals and plants could be readily tested, in contrast to that of human agents. In those early days, therefore, much progress was made on veterinary (and plant) viruses.

Tobacco mosaic virus, TMV, was the first infectious agent identified to be ultrafilterable, by Ivanofsky and Beijerinck (Mayer had earlier shown that the disease could be transmitted by the "juice" of ground leaves of infected plants). However, these studies failed to consider the agent as a novel microorganism. Ivanofsky focused on potential technical aspects of ultracentrifugation, which he assumed had allowed bacteria to pass through the filter. Beijerinck instead did realize that the filterable agent was distinct from bacteria, but he considered the infectivity being in an infectious fluid (from which the name "virus" -poison- was derived). For such reasons, many refer to a virus of cattle, food and mouth disease virus (FMDV), as the first to be discovered. Like Ivanofsky and Beijerinck had found before for TMV, Loeffler and Frosch (in collaboration with Koch) found the agent of FMDV to be ultra-filterable. But Loeffler and Frosch recognized the infection to be transmitted by a novel type of particulate agent much smaller than bacteria. FMDV is therefore often considered as the first virus to be identified. FMDV was without discussion the first mammalian virus identified, and the first virus shown to produce a known infectious disease in organisms other than plants.

Veterinary virology kept on playing a leading role in early virology. While many veterinarian viruses were being discovered, the first human virus, yellow fever virus, was identified only in 1901 and the second one, in 1907. At the end of the first decade after the discovery of FMDV, 13 animal and only 3 human viruses had been identified. Veterinary virology also took the lead in identifying many of the virus families that include the agents of important human diseases. The rabbit myxomavirus was the first poxvirus identified, 22 years before human poxviruses were even observed. Likewise, the swine pseudorabies virus (PRV) was the first herpesvirus identified, 17 years before the first human herpesvirus (herpes simplex type 1). Animal virology also identified the first picornavirus (FMDV), tumor virus, solid tumor virus, influenzavirus, alphavirus, papillomavirus, parvovirus, encephalitis virus, arenavirus, calicivirus and coronavirus, among others. For the first 60 years of virology, there was in fact no discrimination between veterinary and human virology. Such lack of discrimination was in no small part responsible for the explosive growth in virology knowledge during those early years.

Even after the early years, veterinary virology has continued playing major roles in the discovery of viruses that infect humans. In more recent years, the focus has been on the identification of emerging human viruses. Perhaps the most dramatic example is the discovery of Hendra virus as the causative agent of an outbreak that killed 14 horses and their trainer in Australia in 1994 (and half of all other human beings infected since then). A related virus, Nipah, caused a first outbreak in Malaysia in 1998-99, which killed 105 of the 265 people infected. Nipah has continued causing repeated outbreaks ever since, mostly in India and Bangladesh, with a mortality rate of approximately 75%. The latest outbreak was in 2007, but others will likely occur.

Hendra virus was discovered by veterinary virologists at the Animal Health Laboratory of CSIRO Livestock (Australia) as the causative agent of an outbreak of respiratory illness that affected an entire stable and killed 14 racing horses (and their trainer). Following on this work, Nipah was promptly identified as so closely related to Hendra that both viruses are commonly considered together. The group at the Animal Health Laboratory identified the ability of these viruses to infect a variety of domestic and laboratory animals and their routes of secretion. Most importantly from a human health perspective, they further progressed to identify fruit bats ("flying foxes" of the genus Pteropus) as natural reservoirs for these viruses, and horses and pigs as intermediate reservoirs in close proximity to humans. The more recent identification of the mechanisms whereby these viruses enter into cells may eventually result in the development of novel antiviral strategies to prevent infection with such viruses.

The veterinary virologists at the Animal Health Laboratory of CSIRO Livestock have even more recently developed a potential vaccine that protects against these two pathogens. The vaccine showed solid protection of cats against lethal Nipah challenge. Therefore, veterinary virology was critical in identifying these two new human viruses. It was also critical in characterizing their ecology, thus helping to prevent further human infections, and in developing protective measures against infection with these viruses. Veterinary virology therefore continues to play a major role in the identification and characterization of novel viral pathogens that affect human beings.

Novel-viruses are constantly introduced into the human population from their animal reservoirs, as recently shown by SARS and in the more distant past by Junin virus and HIV. In most cases, socio-ecological changes are the most likely cause of the exposure of humans to these new viruses. Deforestation, human settling in remote areas (often to enjoy an "undisturbed" landscape), ecotourism, increases in human population in countries with large previously uninhabited areas, a taste for exotic foods, and climate changes resulting in movement or displacement of wild animals to areas in closer contact with human beings, are all factors contributing to our current enhanced exposure to animal viruses. International travel for business, tourism, or to visit relatives leaving in different countries, is a major factor helping to promptly disseminate any new virus through the world. As none of these factors are likely to drastically change in the near future, our exposure to novel viral pathogens will continue to be exacerbated for the foreseeable future. It is estimated that approximately three quarters of all newly discovered human viruses come from animal reservoirs. Continuous research in veterinary virology is therefore essential to ensure that the required expertise will be available when the next animal virus is introduced into the human population.

Rhabdoviruses

The family Rhabdoviridae includes 20 genera and 144 species of viruses with negative-sense, single-stranded RNA genomes of approximately 10–16 kb. Virions are typically enveloped with bullet-shaped or bacilliform morphology but non-enveloped filamentous virions have also been reported. The genomes are usually (but not always) single RNA molecules with partially complementary termini. Almost all rhabdovirus genomes have 5 genes encoding the structural proteins (N, P, M, G and L); however, many rhabdovirus genomes encode other proteins in additional genes or in alternative open reading frames (ORFs) within the structural protein genes. The family is ecologically diverse with members infecting plants or animals including mammals, birds, reptiles or fish. Rhabdoviruses are also detected in invertebrates, including arthropods some of which may serve as unique hosts or may act as biological vectors for transmission to other animals or plants. Rhabdoviruses include important pathogens of humans, livestock, fish or agricultural crops.

Table: Rhabdoviridae. Characteristics of members of the family Rhabdoviridae.

Characteristic	Description
Typical member	Vesicular stomatitis Indiana virus (AF473864), species *Indiana vesiculovirus*, genus *Vesiculovirus*
Virion	Bullet-shaped or bacilliform particle 100–430 nm in length and 45–100 nm in diameter comprised of a helical nucleocapsid surrounded by a matrix layer and a lipid envelope. Some rhabdoviruses have non-enveloped filamentous virions.

Genome	Negative-sense, single-stranded RNA of 10.8-16.1 kb (unsegmented or bi-segmented).
Replication	Ribonucleoprotein (RNP) complexes containing anti-genomic RNA are generated and serve as templates for synthesis of nascent RNP complexes containing genomic RNA.
Translation	From capped and polyadenylated mRNAs transcribed processively from each gene (3′ to 5′), sometimes containing multiple ORFs.
Host Range	Vertebrates, arthropods and plants; many vertebrate and plant rhabdoviruses are arthropod-borne.
Taxonomy	20 genera containing 143 species and one unassigned species (*Moussa virus*). Many rhabdoviruses remain unclassified.

Viruses assigned to each of the 20 genera form a monophyletic clade based on phylogenetic analysis of L sequences. They usually have similar genome architecture, including the number and locations of accessory genes, and have similarities in host range, modes of transmission and sites of replication in the cell.

Vertebrate Host

Genus *Lyssavirus*: Lyssaviruses infect a wide range of mammals including humans in which they can cause fatal encephalitis (rabies). Natural transmission is via saliva, usually though a bite by an infected animal. The genome is relatively simple, containing the genes which encode five structural protein but feature a long 3′-untranslated region (ψ) in the G gene; additional proteins may be expressed from alternative initiation codons in the P gene.

Genus *Novirhabdovirus*: Novirhabdoviruses infect teleost fish of numerous species in which they can cause severe haemorrhagic disease. Transmission is waterborne and there is also evidence for egg-associated transmission. The genome features an additional gene (NV) that is located between the G gene and L gene. The NV protein appears to be involved in evasion of the host interferon response. Novirhabdoviruses are very distant phylogenetically from fish rhabdoviruses assigned to the genera *Perhabdovirus* and *Sprivivirus*.

Genus *Perhabdovirus*: Perhabdoviruses infect a wide range of teleost fish. They are transmitted through infected water and can cause severe haemorrhagic disease. The genome is relatively simple, containing the five structural protein genes and short intergenic regions. Perhabdoviruses are phylogenetically related to but distinct from fish rhabdoviruses assigned to the genus *Sprivivirus*.

Genus *Sprivivirus*: The viruses assigned to this genus infect a wide range of teleost fish. They are transmitted through infected water and can cause severe haemorrhagic disease. The genome of spriviviruses is relatively simple, containing the five structural protein genes and short intergenic regions. Spriviviruses are phylogenetically related to but distinct from fish rhabdoviruses assigned to the genus *Perhabdovirus*.

Genus *Tupavirus*: Tupaviruses have been isolated from birds, insectivores and rodents, and there is evidence of infection in other vertebrates. The genome features a long alternative ORF in the P gene and an additional gene encoding a small hydrophobic protein between the *M* and *G* genes.

Vertebrate Host and Arthropod Vector

Genus *Curiovirus*: Curioviruses have been isolated from midges, sandflies and mosquitoes. Vertebrate hosts are largely unknown but there is evidence of infection of birds. The genome features one or more genes located between the *M* and *G* genes, and one or more genes located between the *G* and *L* genes, including a gene encoding a viroporin-like protein.

Genus *Ephemerovirus*: Viruses assigned to the genus have been isolated primarily from livestock, mosquitoes or midges. Some cause an acute febrile illness in bovines that is seldom fatal. The genome of ephemeroviruses features multiple genes between the G and L genes encoding accessory proteins including a non-structural class I transmembrane glycoprotein (G_{NS}) and a viroporin (α1).

Genus *Hapavirus*: This genus comprises viruses that have been isolated from mosquitoes or midges and that infect birds and mammals. The genome of hapaviruses is large and complex, featuring multiple accessory genes between *P* and *M* genes, and between *G* and *L* genes, usually including a gene encoding a viroporin-like protein.

Genus *Ledantevirus*: Ledanteviruses infect mammals; many have been isolated from bats or rodents and some (or all) may be transmitted by arthropods. Some have been associated with disease in humans or livestock. The genome is relatively simple but some viruses feature an additional gene between the *G* and *L* genes encoding a small protein of unknown function.

Genus *Sripuvirus*: Viruses assigned to this genus have been isolated from either sandflies or lizards. The genome of sripuviruses features a small protein encoded in a consecutive ORF in the *M* gene and a small transmembrane protein encoded in an alternative ORF at the start of the *G* gene.

Genus *Tibrovirus*: Some tibroviruses infect cattle and water buffalo and are transmitted by midges; several other tibroviruses have been detected in humans but their role in human disease is currently unclear. The genome features two accessory genes between the *M* and *G* genes, and a gene encoding a viroporin-like protein between the *G* and *L* genes.

Genus *Vesiculovirus*: Vesiculoviruses infect a wide range of vertebrate hosts and are transmitted by insects; some may also be transmitted amongst vertebrates by direct contact. Several vesiculoviruses cause vesicular stomatitis in livestock and have been associated with influenza-like illness and occasional encephalitis in humans. The genome is relatively simple, containing the five structural protein genes and short intergenic regions, but may also include alternative ORFs in the *P* gene and use of alternative initiation codons in the *M* gene.

Invertebrate Host

Genus *Almendravirus*: The viruses assigned to this genus were isolated from mosquitoes and appear to be poorly adapted (or not adapted) to replication in vertebrates. The genome of almendraviruses features an additional gene located between the *G* and *L* genes, encoding a small viroporin-like protein.

Genus *Alphanemrhavirus*: This genus comprises viruses that have been detected by high-throughput sequencing in parasitic nematodes (roundworms of the phylum Nematoda). The genome of alphanemrhaviruses is relatively simple, containing the five structural protein genes, but may include an additional small ORF in the *M* gene (Mx) overlapping the end of the M ORF. No alphanemrhaviruses have yet been isolated.

Genus *Caligrhavirus*: Caligrhaviruses have been detected in sea lice (crustaceans in the family Caligidae) in which they appear to cause active infections. The caligrhavirus genome is relatively simple, containing the five structural protein genes, but may include an additional gene (*U1*) between the *G* and *L* genes. No caligrhaviruses have yet been isolated but virions have been observed by electron microscopy.

Genus *Sigmavirus*: Sigmaviruses are transmitted vertically, each virus infecting a fly of a single species in the families Drosophilidae or Muscidae. Infection results in paralysis or death of flies upon exposure to carbon dioxide. The genome may feature an additional gene (*X*) located between the *M* and *G* genes, encoding a protein of unknown function.

Plant Host

Genus *Cytorhabdovirus*: Viruses assigned to this genus infect a wide range of plants and are transmitted by arthropod vectors (aphids, planthoppers and leafhoppers) in which they replicate. In plant cells, cytorhabdoviruses replicate in the cytoplasm. Cytorhabdoviruses have an unsegmented genome featuring an additional gene located between the *P* gene and *M* gene, encoding a movement protein; some may also encode a viroporin-like protein.

Genus *Dichorhavirus*: Dichorhaviruses infect plants and are transmitted by *Brevipalpus* mites. They cause localised lesions on leaves, stems, and fruits of economically important plants such as citrus, coffee and orchids. The genome of dichorhaviruses is bi-segmented: RNA1 contains the *N, P, M* and *G* genes, and an additional gene located between the *P* gene and *M* gene encoding a putative movement protein; RNA2 contains the *L* gene. Virions formed in plant cells may lack envelopes.

Genus *Nucleorhabdovirus*: Nucleorhabdoviruses infect a wide range of plants and are transmitted by arthropod vectors (aphids, planthoppers, leafhoppers) in which they replicate. Nucleorhabdoviruses replicate in the nucleus of infected plant cells. Nucleorhabdoviruses cluster phylogenetically with the bi-segmented dichorhaviruses. They feature an additional gene between the *P* gene and *M* gene encoding a movement protein.

Genus *Varicosavirus*: Varicosaviruses occur naturally in two families of plants (Compositae and Solanaceae) and are transmitted in soil and zoospores of a chytrid fungus, *Olpidium brassicae*. The genome is bi-segmented: RNA1 contains an ORF encoding a small protein followed by the *L* gene; RNA2 contains 5 ORFs including the coat protein gene. Virions observed in plant cells are non-enveloped rods resembling intracellular nucleocapsids of other rhabdoviruses.

Virion

Morphology

Enveloped virions have been reported to be in the range of 100–460 nm in length and 45–100 nm in diameter. The longer forms may represent virions fused end-to-end. Defective-interfering (DI) virus particles are proportionally shorter. Viruses infecting vertebrates are typically bullet-shaped or cone-shaped; however, some rhabdoviruses infecting animals and most plant rhabdoviruses appear bacilliform when fixed prior to staining. In unfixed preparations, they may appear bullet-shaped or pleomorphic. The outer surface of virions (except for the quasi-planar end of bullet-shaped viruses) is covered with projections (peplomers) which are 5–10 nm long and about 3 nm in diameter. They consist of trimers of the viral envelope glycoprotein (G). A honeycomb pattern of peplomers is observed on the surface of some viruses. Internally, the nucleocapsid (30–70 nm in diameter) has helical symmetry and appears to have cross-striations (spacing 4.5–5 nm) in negatively-stained and thin-sectioned virions. The nucleocapsid consists of a ribonucleoprotein (RNP) complex comprising the genomic RNA and tightly bound nucleoprotein (N) together with an RNA-dependent RNA polymerase (L) and polymerase-associated phosphoprotein (P). The RNP complex is active for transcription and replication: the N-RNA template is processed by L, which contains various enzymatic activities, and its cofactor (P). In the cytoplasm, the RNP complex is uncoiled and filamentous, about 700 nm in length and 20 nm in diameter. In the virion, the lipid envelope containing G interacts with the coiled RNP complex via the matrix protein (M). Filamentous virions reported for some plant rhabdoviruses appear to lack a viral envelope.

Rhabdoviridae. (A) Negative-contrast electron micrograph of vesicular stomatitis Indiana virus particles. The bar represents 100 nm. (B) Negative-contrast electron micrograph of RABV defective-interfering (DI) particles. (C) Schematic illustration of a rhabdovirus virion and ribonucleocapsid structure. Unravelling of the RNP is illustrative only to show more clearly its association with the L and P proteins.

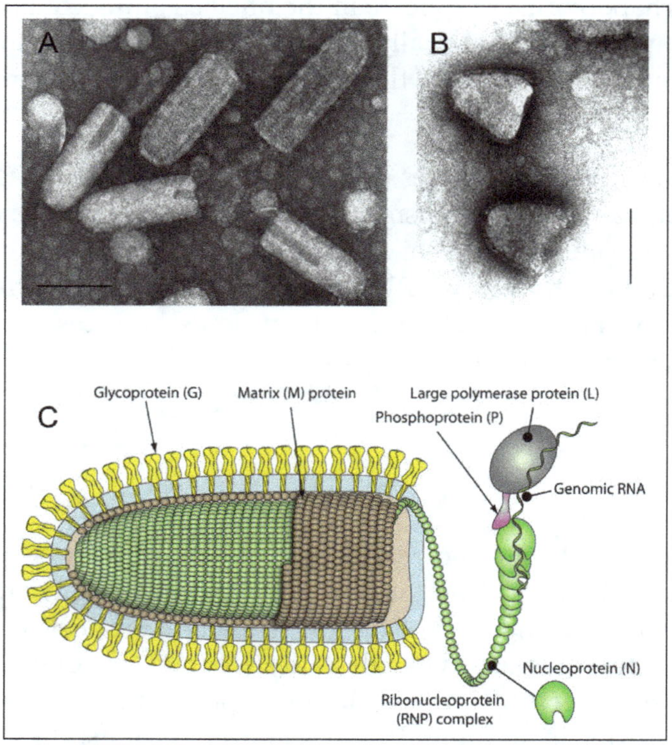

Physicochemical and Physical Properties

Reported virion Mr ranges from 0.3–1.0 x 10^9 and the S_{20w} is in the range 550–1045 S (plant rhabdoviruses usually have larger S_{20w} values). Virion buoyant density is 1.19–1.20 g cm^{-3} in CsCl and 1.16–1.19 g cm^{-3} in sucrose. Virus infectivity is rapidly inactivated at 56 °C, or following UV-, gamma- or X-irradiation, or exposure to formalin or to lipid solvents such as detergents.

Nucleic Acid

Virions typically contain a single molecule of linear, negative-sense single-stranded RNA (Mr 3.4 x 10^6 to 5.4 x10^6; approximately 10–16 kb); rhabdoviruses with segmented genomes also occur with each RNA segment encapsidated independently. The RNA typically represents about 1–3% of virion weight. The RNA has a 3′-terminal free hydroxyl group and a 5′-triphosphate and is not polyadenylated. The ends have inverted complementary sequences encoding transcription and replication initiation signals. Defective-interfering RNAs, usually substantially shorter than full-length RNA (less than half length), may be identified in RNA recovered from virus populations. They are usually negative-sense; however, hairpin RNA forms are also found. Defective-interfering RNAs replicate only in the presence of homologous and, occasionally, certain heterologous helper rhabdoviruses which provide the functional genes. Full-length positive-sense RNA, which is an intermediate during the replication process, may constitute a significant proportion of a viral RNA population. Like the full-length negative-sense RNA genome, the anti-genome is tightly bound to N and does not occur as naked RNA.

Proteins

Virions generally have five structural proteins (designated N, P, M, G and L. The structural proteins represent 65–75% of dry weight of the virion. The function(s) of each of these proteins have been determined largely from studies of the model rhabdoviruses, vesicular stomatitis Indiana virus (VSIV) and rabies virus (RABV); the same functions are typically assumed to apply to other rhabdoviruses, although this is not often confirmed experimentally. Most rhabdoviruses also encode multiple additional (accessory) proteins but few of the encoded proteins have been characterised. Ephemeroviruses express a class 1a viroporin (α1) and proteins with viroporin-like structures occur commonly in animal rhabdoviruses and plant cytorhabdoviruses. Ephemeroviruses and some hapaviruses also express large non-structural class I transmembrane glycoproteins (G_{NS}) that are related to the envelope glycoprotein (G) and appear to have arisen by gene duplication. Novirhabdoviruses infecting fish express a non-structural protein (NV) that appears to be required for efficient replication and plays a role in evading the host innate anti-viral response. Plant-adapted viruses have one or more additional non-structural proteins, one of which has been shown to facilitate virus movement between plant cells. Vesiculovirus express two small proteins (C and C′) from an alternative ORF in the P gene; in lyssaviruses, variant forms of P are expressed from alternative initiation codons in the same frame and are involved in modulating the interferon response.

For certain rhabdoviruses, other nomenclature has previously been used for P (NS, M1 or M2) and M (M1 or M2). The large number and diversity of accessory proteins encoded in rhabdovirus genomes has presented challenges for nomenclature. Some well described accessory proteins have established names that are in common use. However, as the amino acid sequences of most accessory proteins are not highly conserved and their functions are largely unknown, a universal system of nomenclature based on genome location rather than structural or functional homology has been proposed. According to this system: i) each additional transcriptional unit (other than N, P, M, G and L) is designated U (unknown) followed by a number in the order they appear in the genome in positive polarity (i.e., U1, U2, U3, etc); ii) the first ORF within each transcriptional unit is assigned the same designation as the transcriptional unit; and iii) each subsequent ORF (alternative, overlapping or consecutive) within any transcriptional unit is designated with a letter (i.e., U1x, U1y, U1z). Alternative ORFs are defined as those which occur within the frame of a longer ORF; overlapping ORFs are alternative ORFs that extend beyond the frame of the primary ORF; and consecutive ORFs are those which do not overlap but follow consecutively within the same transcriptional unit. The VSIV C and C′ proteins (55 and 65 amino acids, respectively) are the smallest rhabdovirus proteins known to be expressed in infected cells and so ORFs ≥ 180 nucleotides may be considered as potentially significant, depending on their location in the transcriptional unit, the Kozak context of the initiation codon and their conservation in multiple virus isolates or related rhabdoviruses.

Table: *Rhabdoviridae*. Location and functions of rhabdovirus structural proteins.

Protein	Location, mass and function
L	A component of the viral nucleocapsid (ca. 220–240 kDa) responsible for most of the functions required for transcription and replication: RdRP, mRNA 5′-capping, 3′-poly(A) synthesis and protein kinase activities. Observed masses by SDS-PAGE are 150–240 kDa.
G	Associates into trimers to form the virus surface peplomers (monomer ca. 65–90 kDa). Binds to host cell receptor(s), induces virus endocytosis then mediates fusion of viral and endosomal membranes. G is variously N-glycosylated and palmitoylated; it lacks O-linked glycans and may have hemagglutinin activity. Induces and binds virus-neutralizing antibodies and elicits cell-mediated immune responses. In some cases, G is involved in tropism and pathogenicity.
N	Major component of the viral nucleocapsid (ca. 47–62 kDa). It associates with full-length negative- and positive-sense genomic RNAs, or defective-interfering RNAs, but not mRNAs. N is an active element of the template, presenting the bases to the polymerase. Newly synthesised N probably modulates the balance between genome transcription and replication by influencing the recognition of the transcription signals. N elicits cell-mediated immune responses and humoral antibodies. In plant nucleorhabdoviruses, N translocates to a sub-nuclear compartment when co-expressed with the cognate (P).
P	A cofactor of the viral polymerase (ca. 20–30 kDa). It is variously phosphorylated and generally migrates by SDS-PAGE as a protein of about 40–50 kDa; nucleorhabdovirus P migrates faster. P is essential for at least two fundamental functions: (i) it mediates the physical link and the correct positioning of L on the N-RNA template; and (ii) it acts as a chaperone during the synthesis of N, by forming N-P complexes that prevent N from self-aggregation and binding to cellular RNA. During the genome replication process, N is then transferred from these N-P complexes to the nascent viral RNA to ensure its specific encapsidation into new RNPs. P elicits cell-mediated immune responses. In several rhabdoviruses P also plays a fundamental role in evading the host innate anti-viral response.
M	A basic protein that is an inner component of the virion (ca. 20–30 kDa). It is believed to regulate genome RNA transcription. M binds to nucleocapsids and the cytoplasmic domain of G, thereby facilitating the process of budding. It is sometimes phosphorylated or palmitoylated. M is found in the nucleus and inhibits host cell transcription. It also mediates other pathological effects (cell rounding for VSIV, apoptosis for RABV, intracellular accumulation of the inner nuclear membrane for potato yellow dwarf virus (PYDV).

Lipids

Virions are composed of about 15–25% lipid, with their composition reflecting that of the host cell membrane where virions bud. Generally, phospholipids represent about

55–60%, and sterols and glycolipids about 35–40% of the total lipids. G may have covalently associated fatty acids proximal to the lipid envelope.

Carbohydrates

Virions are composed of about 3% carbohydrate by weight. The carbohydrates are present as N-linked glycan chains on G and as glycolipids. Ephemerovirus G_{NS} is also N-glycosylated. In mammalian cells, the oligosaccharide chains are generally of the complex type; in insect cells they are of non-complex types. The number and location of N-glycosylation sites varies for G of different rhabdoviruses.

Genome Organisation and Replication

Rhabdovirus genomes contain at least five ORFs in the negative-sense genome in the order 3′-N-P-M-G-L-5′. The genes are flanked by conserved transcription initiation and termination/polyadenylation signals, about 10 nt in length. For many rhabdoviruses, additional genes are interposed between the structural protein genes and alternative, overlapping or consecutive ORFs may occur within the structural protein genes or in the additional genes. Some rhabdovirus genomes are segmented. Consequently, genomes of viruses assigned to different genera may vary greatly in length and organisation.

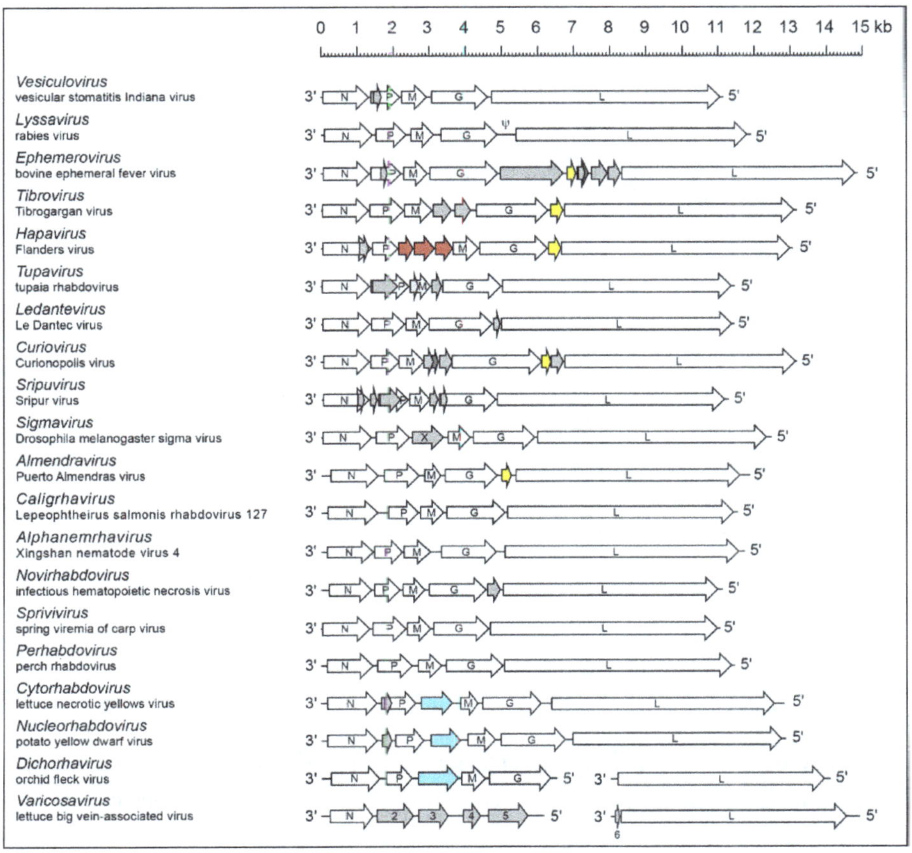

Rhabdoviridae. Schematic representation of rhabdovirus genome organizations exemplifying variations in the number and location of accessory genes. A typical member of each genus is represented. Each arrow indicates the position of a long open reading frame (ORF). Other alternative ORFs occur in some genes; only ORFs (≥180 nt) that appear likely to be expressed are shown.

Most understanding of rhabdovirus replication and transcription has been obtained from studies of vesiculoviruses and lyssaviruses. Genes are transcribed sequentially (from 3′ to 5′ from the template virus RNA and in decreasing molar abundance) as 5′-capped, 3′-polyadenylated, monocistronic mRNAs. A short uncapped, non-polyadenylated and untranslated leader RNA, corresponding to the complement of the 3′-terminus of the viral RNA (i.e., preceding the N mRNA), is also transcribed. Unlike mRNAs, leader RNA has a 5′-triphosphate terminus. Leader RNA of some viruses has been identified in the nucleus of infected cells. The mRNAs generally have common 5′-terminal sequences corresponding to the cap structure fused to the first nucleotides copied from the transcription initiation signal. The mRNAs also each contain a 3′-poly(A) tail which is produced by the viral transcriptase upon copying in a reiterative mode at uridine residues present in each transcription termination signal. Very long 3′-untranslated regions (up to 750 nt) occur in some mRNAs (e.g., lyssaviruses, ephemeroviruses and hapaviruses). Intergenic sequences are generally short but may be up to about 100 nt in length. In some cases, the transcription initiation signal of one gene overlaps the 3′-end of the preceding gene.

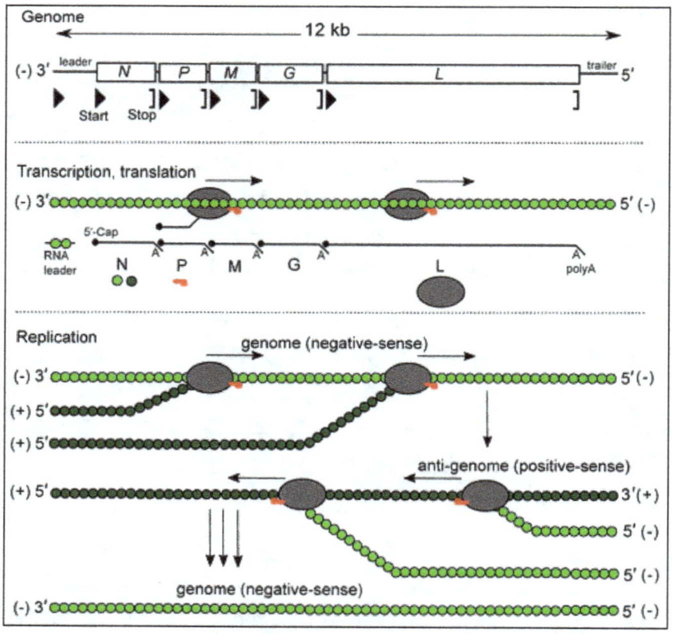

Rhabdoviridae. Genome organization, transcription and replication of vesicular stomatitis Indiana virus. Top: Genome structure Middle: Process of consecutive transcription of leader RNA and messenger RNAs. The role of N (green circles), P (red blob) and

L (grey oval) is indicated. Bottom: replication of the negative-sense genome (light green N) via a positive-sense anti-genome intermediate (dark green N). The switch from transcription to replication is regulated by N. The genome and anti-genome strands are not generated in equimolar amounts.

Non-canonical mechanisms of translation from alternative, overlapping of consecutive ORFs appear to occur commonly in viruses assigned to some genera. Although not yet demonstrated experimentally, the likely mechanisms include: i) leaky ribosomal scanning; ii) a stop-start mechanism involving overlapping or consecutive termination and initiation codons and a 'termination upstream ribosome-binding site' (TURBS); and iii) ribosomal frame shifts featuring a 'slippery' sequence followed by a predicted pseudoknot structure. In the case of some rhabdoviruses, polycistronic mRNAs result from the read-through of the transcription termination signal, allowing transcription extension across the adjacent 5′-gene. However, in most cases, this appears to be due to corruption of the transcription termination signal during adaptation to growth in cell culture.

Except for plant rhabdoviruses, which generally penetrate plant cells through mechanical damage caused by insect vectors, rhabdovirus adsorption is mediated by G attachment to cell surface receptors, and penetration of the cell occurs by endocytosis via coated pits. Various candidate receptors have been postulated for RABV (nicotinic acetylcholine receptor AChR, neural cell adhesion molecule NCAM, low affinity nerve growth factor receptor p75NTR), VSIV (phosphatidyl serine), viral hemorrhagic septicemia virus (VHSV) (fibronectin), and others. In addition, carbohydrate moieties, phospholipids and gangliosides may play a complementary role for virus binding. After penetration by endocytosis, low pH within the endosome triggers fusion between endosomal and viral membranes, liberating the RNP complex into the cytoplasm. The pH-induced fusion depends on conformational changes of the glycoprotein, a process that is reversible upon raising the pH. Once the nucleocapsid is released into the cytoplasm, the RNA genome is repetitively transcribed (primary transcription) by the virion transcriptase. N removal does not occur since the transcriptase only recognizes the RNA-N protein complex as template. The capped and polyadenylated mRNAs are generally translated in cytoplasmic polysomes, except for the G mRNA which is translated on membrane-bound polysomes. Transcription occurs in the presence of protein synthesis inhibitors, indicating that it does not depend on *de novo* host protein synthesis. Following translation, RNA replication occurs in the cytoplasm (full-length positive-sense and then full-length negative-sense RNA synthesis).

Nucleorhabdoviruses and dichorhaviruses replicate in viroplasms in the cell nucleus. Replication again occurs on the RNA-N protein complex and requires the newly synthesised N, P and L species to concomitantly encapsidate the nascent RNA into a nucleocapsid structure. Apart from freshly translated N, P and L, replication may require host factors. Vesiculoviruses can replicate in enucleated cells, indicating that newly synthesised host gene products are not required. However, as for some other negative-sense RNA viruses, trafficking of rabies virus proteins to and from the nucleus appears to play an important role in pathogenesis and modulating the host immune response to infection.

It has been proposed that the concomitant binding of N to the nascent positive- or negative-sense viral RNA species may promote replication rather than transcription, by favoring read-through of transcription termination signals. Replication leads to the synthesis of a full-length positive-sense anti-genome RNA. This, in turn, serves as a replicative intermediate for the synthesis of negative-sense genome RNA for the progeny virions. Following replication, further rounds of transcription (secondary transcription), translation and replication ensue. A typical feature of negative-sense RNA viruses (shared by all members of the order *Mononegavirales*) is that the RNA genome (or anti-genome) is never "naked" in the cell but is always encapsidated by the nucleoprotein. This RNA-N complex is the true template recognised by the viral polymerase (transcriptase or replicase).

Post-translational trafficking and modification of G involves translocation across the endoplasmic reticulum membrane, removal of the amino-proximal signal sequence and step-wise glycosylation in compartments of the Golgi apparatus. Depending on the cell, G may move to the plasma membrane, particularly to the basolateral surfaces of polarised cells.

Viral nucleocapsid structures are assembled in association with M and lipid envelopes containing viral G to form virions. The site of formation of particles depends on the virus and host cell. For vesiculoviruses, lyssaviruses, ephemeroviruses and novirhabdoviruses, nucleocapsids are synthesised in the cytoplasm and virus particles bud from the plasma membrane in most, but not all cells. Some lyssaviruses produce particles that bud predominantly from intracytoplasmic membranes and in some cases prominent virus-specific cytoplasmic inclusion bodies containing N are observed in infected cells (RABV inclusion bodies are called Negri bodies). Cytorhabdovirus virions bud from intracytoplasmic membranes associated with viroplasms; none have been observed to bud from plasma membranes. Nucleorhabdovirus and dichorhavirus virions bud from the inner nuclear membrane and accumulate in the perinuclear space.

Depending on the virus and host cell type, rhabdovirus infections may inhibit cellular protein synthesis and cause apoptosis by mechanisms that are mediated by M. Complementation between viral mutants of related viruses may occur (e.g., between vesiculoviruses), but not between viruses assigned to different genera. Complementation has also been reported to occur by re-utilisation of the structural components of UV-irradiated virus (VSIV). Inter-molecular genetic recombination between different virus isolates is very rare, but intra-molecular recombination may occur during the formation of defective-interfering RNAs. Phenotypic mixing occurs between some animal rhabdoviruses and other enveloped animal viruses (e.g., paramyxoviruses, orthomyxoviruses, retroviruses, herpesviruses).

Antigenicity

G induces virus-neutralising antibodies which define viruses as serotypes and can provide protective immunity. Antigenic cross-reactions in complement-fixation or indirect immunofluorescence tests occur primarily between rhabdoviruses within a genus and involve antigenic determinants located on the N protein. Cross-reactions in indirect immunofluorescence tests have also been detected between some animal rhabdoviruses that are now assigned to different genera.

Biology

Rhabdoviruses are ecologically diverse with members infecting plants or animals including mammals, birds, reptiles or fish. Some of the vertebrate rhabdoviruses have a wide experimental host range; rhabdoviruses infecting plants usually have a narrow host range among higher plants. Rhabdoviruses are also detected in invertebrates, including many arthropods, some of which may serve as biological vectors for transmission to animals or plants. A diverse range of vertebrate and invertebrate cell lines are susceptible to vertebrate rhabdoviruses *in vitro*.

Rhabdoviruses are not usually transmitted vertically in vertebrates or plants, but transovarial transmission has been documented in insects. Sigmaviruses were recognised first as a congenital infection in fruit flies. Vector transmission may involve mosquitoes, sandflies, midges, aphids, leafhoppers or planthoppers. Some viruses are transmitted mechanically in sap or from the body fluids of infected hosts. Mechanical transmission of viruses infecting vertebrates may be by contact, aerosol, bite, or venereal. Fish rhabdoviruses can be transmitted by exposure to infected water.

Genus Demarcation Criteria

Twenty genera have been established to date. Viruses assigned to a genus form a monophyletic clade in well supported Maximum Likelihood trees using full-length L sequences. Use of L for taxonomic purposes is justified by the presence of broadly conserved domains and the rarity of genetic recombination. Demarcation of genera is based upon considerations of significant differences in genome architecture, antigenicity and ecological properties (such as host range, pathobiology and transmission patterns).

Phylogenetic Relationships

Phylogenetic relationships across the family have been established from Maximum Likelihood trees generated from conserved regions of phylogenetically informative sequence in L. These can be identified by aligning full-length L sequences and eliminating ambiguously aligned regions using the Gblocks algorithm. Phylogenetic relationships between viruses assigned to more closely related genera and within genera can also be established using other structural protein genes, notably N and G.

Rhabdoviridae: A Maximum Likelihood phylogenetic tree inferred from a MUSCLE alignment of the full-length L sequences of 134 rhabdoviruses assigned to 20 genera and one rhabdovirus (Moussa virus) representing an unassigned rhabdovirus species. Full-length L sequences are not currently available for other rhabdoviruses assigned to species and so they have not been included in data set. Ambiguously aligned amino acid residues were pruned using Gblocks with 442 positions remaining in the final dataset. The evolutionary history was inferred by using the WAG + frequency model of amino acid substitution with sub-tree pruning and re-grafting (SPR) branch-swapping. The initial tree for the heuristic search was obtained automatically by applying the neighbour-joining algorithm to a matrix of pairwise distances estimated using a JTT model, and then selecting the topology with superior log likelihood value. The tree with the highest log likelihood (-43388.77) is shown. Asterisks (*) indicate well-supported nodes in the tree (bootstrap proportion ≥ 75%) following 1000 iterations. The tree is drawn to scale, with branch lengths measured in the number of substitutions per site, the scale bar indicating a value of 0.5. Evolutionary analyses were conducted in MEGA7.

Similarity with other Taxa

Many general characteristics of rhabdovirus genome organisation, replication and transcription are shared with other members of the order *Mononegavirales*.

Foot-and-mouth Disease Virus

Foot-and-mouth disease (FMD) is a severe, highly contagious viral disease of cattle and swine. It also affects sheep, goats, deer, and other cloven-hooved ruminants. FMD is not recognised as a zoonotic disease.

The disease spreads very quickly if not controlled and because of this is a reportable disease.

Cause

The disease is caused by a virus of which there are seven 'types', each producing the same symptoms, and distinguishable only in the laboratory. Immunity to one type does not protect an animal against other types.

The interval between exposure to infection and the appearance of symptoms varies between twenty-four hours and ten days, or even longer. The average time, under natural conditions, is three to six days.

The virus survives in lymph nodes and bone marrow at neutral pH, but is destroyed in muscle when pH is less than 6.0, i.e., after rigor mortis. The virus can persist in contaminated fodder and the environment for up to one month, depending on the temperature and pH conditions.

Airborne spread of the disease can take place and under favourable weather conditions the disease may be spread considerable distances by this route.

Animals pick up the virus either by direct contact with an infected animal or by contact with foodstuffs or other things which have been contaminated by such an animal, or by eating or coming into contact with some part of an infected carcase. Outbreaks have been linked with the importation of infected meat and meat products.

The disease can also be spread by people, vehicles and other objects that have been contaminated by the virus.

Symptoms

- Fever.
- Bilsters in the mouth and on feet.
- Drop in milk production.
- Weight loss.
- Loss of appetite.
- Quivering lips and frothing of mouth.
- Cows may develop blisters on teats.
- Lameness.

Treatment

Treatment is not given. Affected animals will recover. However because of the loss of production and the infectious state of the disease, infected animals are usually culled.

Prevention

FMD is one of the most difficult animal infections to control. Because the disease occurs in many parts of the world, there is always a chance of its accidental introduction into an unaffected country. Export restrictions are often imposed on countries with known outbreaks.

FMD outbreaks are usually controlled by quarantines and movement restrictions, euthanasia of affected and in-contact animals, and cleansing and disinfection of affected premises, equipment and vehicles.

Infected carcasses must be disposed of safely by incineration, rendering, burial or other techniques. Milk from infected cows can be inactivated by heating to 100 °C (212 °F) for more than 20 minutes. Slurry can be heated to 67 °C (153 °F) for three minutes.

Rodents and other vectors may be killed to prevent them from mechanically disseminating the virus.

Good biosecurity measures should be practiced on uninfected farms to prevent entry of the virus.

Vaccination

- Vaccination can be used to reduce the spread of FMD or protect specific animals.
- Vaccines are also used in endemic regions to protect animals from clinical disease. FMDV vaccines must closely match the serotype and strain of the infecting strain.
- Vaccination with one serotype does not protect the animal against other serotypes, and may not protect the animal completely or at all from other strains of the same serotype. Currently, there is no universal FMD vaccine.

Pestiviruses

Pestivirus (also known as BVDV - bovine viral diarrhoea virus) causes abortion, ill thrift in young animals, diarrhoea and respiratory disease.

Transmission is by direct contact with a carrier animal. The virus is common in cattle and many herds are infected.

Conditions when Pestivirus is likely to Occur

- Close contact between cattle.
- Recent introduction of a carrier animal(s) into a herd of susceptible pregnant females.
- Introduction of new cattle into a closed breeding herd.
- Introduction of new cattle into a breeding herd in the early stages of pregnancy.
- When the breeding herd has access to other cattle on the property.

Identifying and Diagnosing Pestivirus

Diagnosis of pestivirus will require veterinary assistance. Clinical signs of pestivirus can vary depending on the strain of virus and time of infection.

Clinical signs that would lead a producer to suspect pestivirus include:

- Early-term abortion or embryonic loss.
- Temporary infertility.
- Increased susceptibility to other diseases.
- Weak, stunted or deformed calves.
- Diarrhoea.
- Respiratory disease.
- Ill-thrift and wastage.

Prevention Strategies for Pestivirus

- Defining the pestivirus status of a herd by serological testing of herds.
- Identifying and culling persistently infected animals in conjunction with a veterinarian.
- Only buying cattle from other uninfected properties.
- Purchasing cattle from properties with no history of trading, agistment or cattle turnover, compared to cattle trading properties or where agistment is run.
- Keeping newly purchased cattle away from the breeding herd, especially if in early pregnancy.
- Ensuring replacement females have developed a strong immunity before joining.
- Vaccinating to control the disease in cattle.

Arteriviruses

Arteriviruses are pleomorphic but roughly spherical particles. By cryo-electron microscopy porcine reproductive and respiratory syndrome virus (PRRSV) particle diameters were found to range from 50 to 74 nm, with a median value of 54 nm and only few particles larger than 60 nm. Using the same approach the average diameter of the isometric nucleocapsid, which is probably not icosahedrally ordered, was found to be 39 nm. The nucleocapsid is surrounded by a lipid envelope with small surface projections that cover the entire virion surface.

Physicochemical and Physical Properties

The buoyant density of arterivirus particles has been estimated to be 1.13 to 1.17 g cm^{-3} in sucrose. Reported sedimentation coefficients for arteriviruses range from 200S to 300S. Virions are stable when stored at −70 °C. The half-life of arteriviruses progressively decreases with increasing temperature. Virions are stable between pH 6.0 and 7.5, but are inactivated at high or low pH. Arteriviruses are also inactivated by lipid solvents, such as ether, butanol and chloroform and are extremely sensitive to detergent treatment. A brief incubation with a nonionic detergent such as 0.01% NP40 or Triton X-100 efficiently disrupts the viral envelope.

Nucleic Acid

Virions contain a single molecule of linear, positive sense, single stranded RNA that ranges in length from 12.7 to 15.7 kb. The naked RNA, when transfected into permissive cells, is itself infectious. The genomic RNA contains a 5′ type I cap structure (simian hemorrhagic fever virus, SHFV) and a 3′-terminal poly(A) tract. Full-length sequences are available in the GenBank database for representatives of all currently known arterivirus species.

Proteins

Seven structural proteins have been identified in equine arteritis virus (EAV) and PRRSV virions. In addition to the nucleocapsid protein (N), there are two major (GP5 and M) and four minor (E, GP2, GP3, GP4) envelope proteins. By reverse genetics (EAV and PRRSV), each of these proteins was shown to be required for the production of infectious progeny. The major glycoprotein, GP5, spans the membrane three times and forms a disulfide-linked heterodimer with triple membrane spanning M protein; the heterodimer between conserved cysteine residues is essential for virus infectivity. The GP5 proteins of EAV and SHFV are predicted to possess 98 residues on the outside of the virion, while both genotypes of PRRSV and lactate dehydrogenase-elevating virus (LDV) are predicted to have ectodomains of approximately 30 amino acids. The predicted orientation of the minor envelope proteins in the viral membrane is also shown in figure. The E protein of EAV and PRRSV is fatty-acid acylated, and the myristoylation of E has been shown to be non-essential for virus infectivity in both viruses. The E protein appears to be an ion-channel protein that may function in the uncoating process during virus entry and penetration. GP2, GP3 and GP4 form heterotrimers on the surface of the virus particle (EAV and PRRSV). The proteins encoded by ORFs 2a/b, 3 and 4 have not been confirmed as structural components of LDV, nor has the trimerization of these proteins been assessed. A soluble, non-virion associated form of the ORF3 glycoprotein is also released from infected cells (LDV and PRRSV Type 2). In addition, the N proteins of EAV and PRRSV Type 2 have been shown to dimerize. Crystal structures of the putative dimerization domain of arterivirus N proteins (PRRSV Type 2 and EAV) have been determined and indicate that these proteins represent a new class of viral capsid-forming proteins. Virus mutants (EAV) lacking expression of E, GP2, GP3 or GP4 produce non-infectious particles. The virion proteins of SHFV include the two major envelope proteins (GP5 and M) and the N protein; the genome possesses four additional 3′ ORFs (2a′, 2b′, 3′ and 4′), encoding GP2′, E′, GP3′ and GP4′, which may be duplications of ORFs 2 a/b to 4.

Table: Virion-associated proteins of arteriviruses.

Protein	aa[b]	ORF	mRNA	(Putative) Function(s)
E[d]	67–80	2a/2b[c]	2[d]	Small integral envelope protein, myristoylated, postulated ion channel protein.
GP2a/b	227–256	2b/2a[c]	2[d]	Minor glycoprotein, part of GP2/GP3/GP4 heterotrimer.
GP3	163–265	3	3	Minor glycoprotein, part of GP2/GP3/GP4 heterotrimer.
GP4	152–183	4	4	Minor glycoprotein, part of GP2/GP3/GP4 heterotrimer.
GP5	199–278	5	5	Major glycoprotein, carries main determinants for neutralization, part of GP5/M heterodimer.
M	162–174	6	6	Integral membrane protein, part of GP5/M heterodimer.
N[e,f]	110–128	7	7	Nucleocapsid protein, partially localizes to the nucleus of infected cells, phosphoprotein.

a - Based on the genome organization of EAV, PRRSV and LDV, adapted from the Eighth ICTV Report; SHFV ORFs 2a-7 were used to predict structural protein amino acid length.

b - aa=amino acids.

c - In PRRSV.

d - Subgenomic mRNA2 is assumed to be functionally bicistronic.

e - Partial crystal structures have been determined for EAV and PRRSV N (Protein Data Bank IDs 2I9F and 1P65, respectively).

f - ORF7 polymorphism in Type 1 (125, 126, 129, 131 AA) and Type 2 (124, 125 AA) PRRSV.

Lipids

The virion lipids are cell derived, with virus budding occurring on membranes of the ER and the Golgi part of the exocytic pathway.

Carbohydrates

Although the number of predicted N-linked glycosylation sites in GP2-5 of EAV, PRRSV and LDV strains differ, the large majority of these sites are highly conserved and present in almost all natural isolates of the three viruses. Specifically, it has been demonstrated that conserved N-linked glycosylation sites in GP5 are essential for virus infectivity and antigenicity (EAV and PRRSV). Recent reverse genetics studies have revealed that PRRSV and EAV mutants void of GP5 N-linked glycosylation are nonviable or readily revert to restore at least one N-linked glycosylation site. Since stable virus mutants lacking these carbohydrates cannot be generated, further investigation of the specific role of glycans is difficult. However, a viable pseudorevertant (deletion mutant) of an EAV mutant with a completely unglycosylated GP5 ectodomain was isolated. This pseudorevertant grew to a lower titer and produced smaller plaques compared to the parental virus with a single conserved GP5 glycosylation site. The sugar composition of the N-linked glycans of GP2-5 of these arteriviruses has not been investigated in detail. Preliminary studies indicate that the EAV GP5 carbohydrate is modified with N-acetyllactosamine, while PRRSV GP5 is thought to contain complex sugars other than polylactosaminoglycans.

Antigenic Properties

Each of the four arteriviruses is antigenically distinct and arteriviruses show no serological cross-reactivity. Furthermore, there is significant antigenic variation among different strains of EAV, LDV and PRRSV, but the antigenic properties of SHFV have not been evaluated. The major neutralization epitopes of arteriviruses have been mapped to the GP5 protein of EAV, PRRSV (both Type 1 and 2) and LDV and SHFV 5. It has been shown that GP5 and M heterodimerization is critical for the expression of neutralization epitopes on GP5 proteins of EAV and PRRSV. In contrast to Type 2 PRRSV, neutralizing antibodies against GP4 have also been shown to neutralize the Type 1 PRRSV prototype (Lelystad). In addition to GP5, the M protein is the most frequently recognized EAV structural protein by convalescent sera from non-persistently infected horses, whereas sera from carrier stallions also recognize the GP3 and N proteins. The antigenicity of and humoral antibody response to EAV nsps have yet to be determined. Antibodies to several nonstructural proteins (nsps; nsp1, nsp2 and nsp7), N and M proteins and GP5 appear within 7 to 14 days of PRRSV infection. However, the neutralizing antibody response against GP5 is delayed. A number of B cell epitopes have been identified in the nsp2 protein of PRRSV. Similarly, a number of T-cell epitopes have been identified on GP4, GP5, M and N proteins of PRRSV. Infection with PRRSV also leads to suppression of T lymphocytes, and a number of nsps have been implicated in countering host immune responses. Infection with some PRRSV strains result in a transient induction of high levels of serum interferon gamma. Mice infected with LDV develop a strong neutralizing antibody response to GP5 early in the infection that effectively neutralizes the virus. Nevertheless, virus isolated from persistently infected mice is neutralization-resistant, suggesting selection of neutralization escape variants. The activation of T lymphocytes by LDV infection is limited to the first day postinfection and is triggered by the large amounts of interferon alpha produced by the initial productively-infected macrophages. This T-lymphocyte activation is rapidly followed by a transient suppression of T cell responses (cytotoxic and helper) that is maximal at about 3 days postinfection. LDV and PRRSV (in gnotobiotic pigs) also trigger polyclonal B-cell activation.

Arterivirus Entry

The arterivirus cell tropism has been attributed to the presence or absence of specific entry mediators. The entry of PRRSV into the porcine macrophage has been intensively studied but little is yet know about the entry of EAV, LDV and SHFV into their respective target cells. PRRSV virions adhere to macrophages through interactions with heparan sulphate glycosaminoglycans that line the cell surface. Subsequently, the virus binds to the macrophage-specific lectin sialoadhesin. Sialic acids on the virion surface are crucial for this interaction and the M/GP5 glycoprotein complex was identified as a ligand for this lectin receptor. Binding of PRRSV to sialoadhesin triggers uptake of the virus-receptor complex via clathrin-mediated endocytosis and subsequent release of the viral genome into the cytoplasm of the target cell initiates the translational and transcriptional processes that lead to productive infection. PRRSV genome release is dependent on acidification of the virus-containing endosome, and both cellular CD163 and proteases (cathepsin E) appear to play a crucial role in this step. GP2 and GP4 were identified as binding partners for CD163 by co-precipitation experiments, indicating a potential function for these molecules in genome release.

Genome Organization and Replication

For EAV, PRRSV and LDV, the genome contains nine functional ORFs, whereas the single reported SHFV sequence contains 13 ORFs, due to an apparent 4-gene duplication. The viral genes for EAV, PRRSV and LDV are arranged in the order 5′-replicase-E/GP2-GP3-GP4-GP5-M-N-3′, and mostly overlap. A schematic of the arterivirus replication cycle, based on studies with EAV, is shown in figure. Following genome translation, dedicated nonstructural proteins (nsps) are thought to induce double-membrane vesicles (DMVs) with which the replication/transcription complexes of arteriviruses become associated in order to engage in RNA-dependent RNA synthesis. In addition to a full-length minus strand, infected cells also contain a nested set of subgenomic minus strand RNAs that are the complements of the subgenomic mRNAs (sgmRNAs) and are believed to function as templates for their transcription. The 5′-proximal 156–211 nt of the sgmRNAs ("leader" or 5′ UTR) are derived from the 5′ end of the genome, whereas the coding region of the sgmRNA (the mRNA "body") is colinear with the 3′-proximal region of the genome. The synthesis of the subgenomic minus strands, which involves the fusion of sequences that are noncontiguous in the genome, is currently thought to occur by discontinuous extension during minus strand synthesis. This is a process in which transcription-regulating sequences (TRSs) direct attenuation of minus strand synthesis, translocate the nascent strand to the leader region in the 5′ end of the genomic template, and (guided by a base-pairing interaction) reinitiate minus strand synthesis to add the leader complement to the nascent subgenomic minus strand RNA.

High frequency intraspecies RNA recombination between divergent strains of EAV, LDV and PRRSV has been conclusively shown. For PRRSV, only intragenotype (Type 1 or 2) recombination has been detected to date.

Four cell proteins (36, 55, 86 and 103 kDa) have been identified that bind to a *cis*-acting region required for plus-strand RNA synthesis from the minus-strand template, and two cell proteins (polypyrimidine tract-binding protein and fructose 1,6 bisphosphate aldolase) have been shown to bind to the 3′ UTR of the positive strand, required for negative strand synthesis of SHFV and LDV-C. Other cell proteins have been identified that bind to the 5′ UTRs of EAV positive and negative strand RNAs.

Nonstructural Proteins

The arterivirus genome is polycistronic and contains 9 to 13 ORFs. Three-quarters of the 5'-proximal length of the genome is occupied by two large ORFs that together encode all viral enzyme functions (collectively referred to as the viral "replicase", figure) required for genome replication and subgenomic mRNA production. Replicase ORFs 1a and 1b are translated into polyproteins pp1a (187–260 kDa) and pp1ab (345–421 kDa), with the latter being a C-terminally extended version of the former. ORF1b translation depends on a -1 ribosomal frameshift just before termination of ORF1a translation. A number of conserved domains are present in the replicases of all arteriviruses (from N-terminus to C-terminus; figure): a zinc finger (ZF in nsp1), a papain-like cysteine proteinase (PLP1α and β [and PLP1γ in SHFV) in nsp1], an unusual papain-like cysteine proteinase (PLP2 or CP in nsp2), a chymotrypsin-like serine proteinase (SP in nsp4; alternatively known as serine Main protease, M^{pro}), a RNA-dependent RNA polymerase (RdRp in nsp9), a zinc binding domain (Z in nsp 10), a NTPase/RNA helicase (HEL in nsp10), and a nidovirus uridylate-specific endoribonuclease (NendoU; U in nsp11).

The RNA polymerase (EAV), helicase (EAV and PRRSV Type 2), and endoribonuclease activities (EAV and PRRSV Type 2) of nsp9, nsp10 and nsp11, respectively, have been corroborated with biochemical assays using purified recombinant proteins. Furthermore, EAV nsp1 has been identified as a critical regulator of the accumulation levels of genomic and subgenomic mRNAs, most likely by modulating the synthesis of their corresponding minus strand templates. ORF1a-encoded subunits with hydrophobic domains, in particular nsp2 and nsp3, have been implicated in the formation of the endoplasmic reticulum-derived DMVs with which the viral replication and transcription complexes are associated. Detailed analysis has shown that the EAV replicase polyproteins pp1a and pp1ab are cleaved into 13 mature nsps by the three viral proteinases (PLP1β, PLP2 and SP), whereas LDV and PRRSV produce at least one additional cleavage product due to the fact that their nsp1 equivalent contains a second internal proteinase that cleaves the nsp1 region into nsp1α and nsp1β. PLP2 (CP) has been shown to cleave nsp2 from nsp3 in EAV and PRRSV, and to assist the SP in cleavage of the nsp4/5 junction in EAV. The EAV nsp4 SP is responsible for nine proteolytic cleavages in the C-terminal half of pp1a and the ORF1b-encoded part of pp1ab. Several cysteine/histidine (C/H) motifs have been identified, and two (putative) zinc binding domains (in nsp1 and nsp 10) have been shown to be critical for the EAV replication cycle. By sequence comparison, similar cleavage sites and C/H motifs have been predicted in all other arteriviruses, but most are not confirmed. The PLP1β cleavage site for PRRSV has been shown to be different from what was originally proposed.

Table: Nonstructural proteins of EAV.

Protein	aa[b]	Mode of expression[c]	(Putative) Function(s)
nsp1[d]	260	TI + nsp1 PLP1β	Zinc finger, proteinase (PLP1), replicase polyprotein processing, transcription, and virion biogenesis (dispensable for genome replication).
nsp2	571	TI + nsp1 PLP1β + nsp2 PLP2	Proteinase (PLP2) with deubiquitinating activity (DUB), integral membrane protein, replication complex (DMV) formation.
nsp3	233	TI + nsp2 PLP2 + nsp4 SP	Integral membrane protein, replication complex (DMV) formation.
nsp4[e]	204	TI + nsp4 SP	Main proteinase (SP).
nsp5	162	TI + nsp4 SP	Integral membrane protein, replication complex (DMV) formation.
nsp6	22	TI + nsp4 SP	?
nsp7α, β	225	TI + nsp4 SP	?
nsp8[f]	50	TI + nsp4 SP + TT	?
nsp9	693	TI + RFS + nsp4 SP	RNA-dependent RNA polymerase.
nsp10	467	TI + RFS + nsp4 SP	RNA helicase/NTPase, putative zinc binding domain, role in subgenomic mRNA synthesis.
nsp11	219	TI + RFS + nsp4 SP	Nidovirus uridylate-specific endoribonuclease.
nsp12	119	TI + RFS + nsp4 SP + TT	?

a - Based on the currently known replicase processing scheme of EAV, adapted from the Eighth ICTV Report.

b - aa=amino acids.

c - TI, translation initiation; RFS, ORF1a/ORF1b ribosomal frameshifting; TT, translation termination; PLP1, papain-like cysteine proteinase; PLP2, papain-like, cysteine proteinase; SP, serine proteinase.

d - Nsp1 of LDV and PRRSV is cleaved internally by an additional papain-like proteinase to yield nsp1α and nsp1β. The nsp1α subunit of PRRSV contains two distinct zinc finger configurations, the second of which has been shown to be functional. A crystal structure has been determined for PRRSV nsp1α (Protein Data Bank ID 3IFU).

e - Crystal structures have been determined for EAV and PRRSV nsp4 (Protein Data Bank IDs 1MBM and 3FAO, respectively).

f - Due to ribosomal frameshifting, nsp8 is identical to the N-terminal 50 aa of nsp9.

Biological Properties

All known arteriviruses infect a single type of vertebrate host in the domain Eucarya and are not transmitted by a vector. Macrophages are the primary host cell. Arteriviruses are cytocidal and usually cause both acute and chronic, persistent disease in their hosts. EAV is the agent of equine viral arteritis (inflammation of small arteries) and infection can lead to extremely variable clinical signs. Viral arteritis has not been a reported characteristic of infection with LDV and PRRSV. LDV causes a lifelong persistent infection of mice, but CNS disease resulting in paralysis is seen in Fv-1n/n mice. Infection of swine with PRRSV results in reproductive failure in sows, respiratory illness in growing swine, and is usually asymptomatic in boars. For PRRSV, extraordinary strain diversity has been detected and is due in part to high levels of viral recombination (Type 1 and 2) that leads to variation in clinical symptoms from mild to severe (especially in Type 2). SHFV causes acute or persistent infections in African monkeys, such as patas (*Erythrocebus patas*), with no overt disease symptoms, but induces a fatal hemorrhagic fever in monkeys of the genus *Macaca*.

Species Demarcation Criteria in the Genus

Each of the four currently recognized species in the genus *Arterivirus* constitutes a major phylogenetic branch within the genus. For each species except SHFV, multiple full-length sequences were determined and indicate that the species consist of clusters of sometimes widely divergent strains. LDV and PRRSV are most closely related to each other. There are two subspecific types of PRRSV, European (Type 1) and North American (Type 2). Members of the four virus species are antigenically distinct. The viruses of each species have a restricted host range and each species infects different hosts. The size variation of the N-terminal half of ORF1 suggests that this arterivirus region may encode species-specific functions.

Coronaviruses

Coronaviruses are types of viruses that typically affect the respiratory tract of mammals, including humans. They are associated with the common cold, pneumonia, and severe acute respiratory syndrome (SARS) and can also affect the gut.

A coronavirus was first isolated in 1937 from an infectious bronchitis virus in birds that has the ability to seriously devastate poultry stocks. These viruses are responsible for between 15 and 30 percent of common colds.

Over the last 70 years, scientists have found that coronaviruses can infect mice, rats, dogs, cats, turkeys, horses, pigs, and cattle.

This MNT Knowledge Center article will focus on the different types of human coronaviruses, their symptoms, how they are transmitted, and two particularly dangerous diseases that can be caused by coronaviruses: SARS and MERS.

Fast Facts on Coronaviruses

- There is no cure for the common cold.
- A coronavirus causes both SARS and MERS.
- Coronaviruses infect many different species.
- There are six known human coronaviruses.
- SARS spread from China to cause infection in 37 countries, killing 774 people.

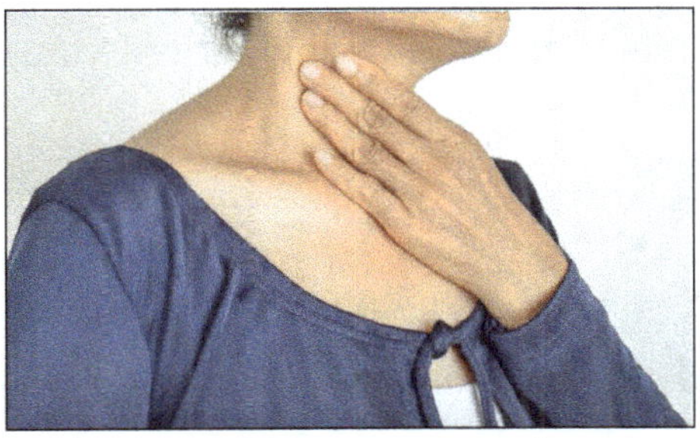

Coronaviruses can cause flu-like symptoms and respiratory symptoms.

Human coronaviruses (HCoV) were first identified in the 1960s in the noses of patients with the common cold. Two human coronaviruses are responsible for a large proportion of common colds OC43 and 229E.

Coronaviruses were given their name based on the crown-like projections on their surfaces. "Corona" in Latin means "halo" or "crown."

Among humans, infection most often occurs during the winter months as well as early spring. It is not uncommon for a person to become ill with a cold that is caused by a coronavirus and then catch it again about four months later.

This is because coronavirus antibodies do not last for a very long time. Also, the antibodies for one strain of coronavirus may be useless against other strains.

Symptoms

Cold- or flu-like symptoms usually set in from two to four days after coronavirus infection, and they are typically mild.

Symptoms include:

- Sneezing.
- A runny nose.
- fatigue.
- A cough.
- In rare cases, fever.
- A sore throat.
- Exacerbated asthma.

Human coronaviruses cannot be cultivated in the laboratory easily, unlike the rhinovirus, another cause of the common cold. This makes it difficult to gauge the coronavirus' impact on national economies and public health.

There is no cure, so treatments include taking care of yourself and over-the-counter (OTC) medication:

- Rest and avoid overexertion.
- Drink enough water.
- Avoid smoking and smoky areas.
- Take acetaminophen, ibuprofen or naproxen to reduce pain and fever.
- Use a clean humidifier or cool mist vaporizer.

The virus responsible can be diagnosed by taking a sample of respiratory fluids, such as mucus from the nose, or blood.

Types

Different types of human coronaviruses vary in the severity of illness they cause and how far they can spread.

There are currently six recognized types of coronavirus that can infect humans. Common types include:

- 229E (alpha coronavirus).
- NL63 (alpha coronavirus).
- OC43 (beta coronavirus).
- HKU1 (beta coronavirus).

Rarer, more dangerous types include MERS-CoV, which causes Middle East Respiratory Syndrome (MERS), and severe acute respiratory syndrome (SARS-CoV), the coronavirus responsible for SARS.

Transmission

Contagious diseases can spread through coughing without covering the mouth.

There has not been a great deal of research on how a human coronavirus spreads from one person to the next.

However, it is believed that the viruses transmit using secreted fluid from the respiratory system.

Coronaviruses can spread in the following ways:

- Coughing and sneezing without covering the mouth can disperse droplets into the air, spreading the virus.
- Touching or shaking hands with a person that has the virus can pass the virus from one person to another.
- Making contact with a surface or object that has the virus and then touching your nose, eyes, or mouth.
- On rare occasions, a coronavirus may spread through contact with feces.

People in the U.S. are more likely to contract the disease in the winter or fall. The disease is still active during the rest of the year. Young people are most likely to contract a coronavirus, and people can contract more than one infection over the course of a lifetime. Most people will become infected with at least one coronavirus in their life. It is said that the mutating abilities of the coronavirus are what make it so contagious.

To prevent transmission, be sure to stay at home and rest while experiencing symptoms and avoid close contact with other people. Covering the mouth and nose with a tissue or handkerchief while coughing or sneezing can also help prevent the spread of a coronavirus. Be sure to dispose of any used tissues and maintain hygiene around the home.

SARS

Severe Acute Respiratory Syndrome (SARS) was a contagious disease caused by the SARS-CoV coronavirus. It typically led to a life-threatening form of pneumonia.

The virus started off in the Guangdong Province in southern China in November 2002, eventually reaching Hong Kong. From there, it rapidly spread around the world, infecting people in 37 countries.

SARS-CoV is unique. It can infect both the upper and lower respiratory tract and can also cause gastroenteritis.

The symptoms of SARS develop over the course of a week and start with a fever. Early on in the condition, people develop flu-like symptoms, such as:

- Dry coughing.
- Chills.
- Diarrhea.
- Breathlessness.
- Aches.

Pneumonia, a severe lung infection, may develop afterward. At its most advanced stage, SARS causes failure of the lungs, heart, or liver.

During the epidemic, there were 8,098 confirmed cases of SARS with 774 fatalities. This is equal to a mortality rate of 9.6 percent. Complications were more likely in older adults, and half of all infected people over the age of 65 years who became ill did not survive. It was eventually brought under control in July 2003.

MERS

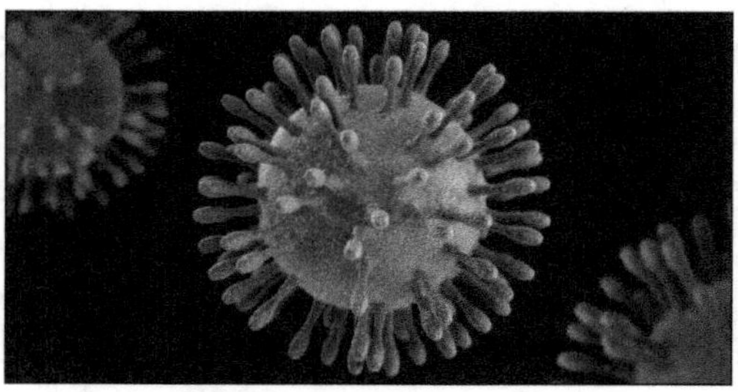

MERS is a potentially fatal coronavirus.

MERS, caused by the MERS-CoV coronavirus, was first recognized in 2012. This severe respiratory illness first surfaced in Saudi Arabia and, since then, has spread to other countries. The virus has reached the U.S., and the largest outbreak outside the Arabian Peninsula occurred in South Korea in 2015.

Symptoms include fever, breathlessness, and coughing. The illness spreads through close contact with people who have already been infected. However, all cases of MERS are linked to individuals who have recently returned from travel to the Arabian Peninsula. MERS is fatal in 30 to 40 percent of people who contract it.

Influenza

Influenza is a respiratory illness common to humans and a limited number of animal species – namely domestic and wild birds, pigs, wild aquatic mammals (seals and whales), minks and farmed carnivores. Among the three types of influenza viruses, type A virus is the important one as far as cross-species infections are concerned.

Although there is certain degree of host-specificity among influenza viruses, barriers between species are by no means absolute. It is, thus, of utter most importance to understand the zoonotic potential of certain animal influenza viruses and the role of animals in the generation of novel human influenza viruses. Appreciation of the zoonotic nature of influenza viruses and adequate surveillance of animal populations are pivotal steps in tackling future influenza pandemics.

Examples of Zoonotic Influenza

Influenza viruses were initially isolated from pigs in the United States in the early 1930s and remained one of the most frequently encountered swine respiratory diseases. The isolates were of the H1N1 subtype, and their characterization followed the deadly Spanish flu in 1918. To this day the question remains open whether these viruses first appeared in people and then spread to pigs, or vice versa.

In pigs, influenza most commonly presents in the form of explosive outbreaks of acute respiratory disease characterized by fever and lethargy, nasal and ocular discharge, coughing, dyspnea and loss of appetite. The disease lasts only for a week, usually with full recovery. Although mortality is low, swine influenza may cause an economic loss due to the delayed weight gain.

The nature of influenza virus infections in birds is reliant upon two major factors – the species of the affected bird and the pathogenicity of the virus. The causative viruses are of the H5 and H7 subtypes. Clinical disease develops in chickens and turkeys, while infections in ducks and most other waterfowl are subclinical, i.e. without any apparent symptoms.

Lethargy, loss of appetite, weight loss, ruffled feathers and reductions in egg production are symptoms that follow infection with avian influenza of relatively low pathogenicity, making the diagnosis in the field difficult. On the other hand, highly virulent strains cause fatal systemic disease ("fowl plague") characterized by edema, hemorrhages and infection of the central nervous system.

Equine influenza is caused by two subtypes of influenza A viruses - H7N7 and H3N8. Clinical signs include extreme fever, nasal discharge and a harsh dry cough; rarely pneumonia in young foals and donkeys and encephalitis in horses can develop.

In 2004, cases of an unknown respiratory illness in dogs started to emerge. An investigation showed that the disease was caused by the equine influenza A H3N8 virus, so experts believe that it "jumped" from horses. Symptoms are also similar to equine influenza, and approximately 80 percent of infected dogs have a mild form of the disease.

Transmission from Animals to People

Wild birds represent the primary natural reservoir for all subtypes of influenza A viruses and are considered as a source of influenza A viruses in all other animals. In addition to swine influenza viruses, pigs can be infected with both human and avian influenza viruses.

Because of such susceptibility to the variety of influenza viruses, pigs are often considered as intermediary hosts or mixing vessels. They can be infected with influenza viruses from different species at the same time, thus creating a milieu for the genes of these viruses to mix and generate a new virus. This event is called reassortment.

Such new virus could adapt to the mammalian host and spread from person to person, but its surface proteins (hemagglutinin and neuraminidase) would not be previously seen in influenza viruses that infect humans. This type of abrupt, major change in the influenza A viruses is known as antigenic shift, and it can result in a worldwide influenza pandemic.

Highly pathogenic avian influenza A (H5N1) virus (often referred to as HPAI H5N1) is a virus that occurs predominantly in birds, transmits easily in these animals and can be deadly to them, especially to domestic poultry. Even though they are relatively rare, sporadic human infections with this virus have occurred and caused significant morbidity and mortality. Although there is no evidence suggesting efficient human-to-human transmission of HPAI H5N1, researching this virus is one of the top priorities.

Thus far there is no evidence of canine influenza virus transmission from dogs to people and no reported human infections. As close contact between animals, stress, environmental and meteorological factors have been shown to contribute to influenza virus spread, adequate public health measures and disease awareness remain critical.

Bluetongue Virus

Bluetongue virus (BTV), an economically important orbivirus of the Reoviridae family, causes a haemorrhagic disease mainly in sheep and occasionally in cattle and some species of deer. BTV is transmitted between its mammalian hosts by certain species of biting midges (Culicoides spp. – Diptera: Ceratopogonidae) and it can infect all ruminant species. Although BTV is an arbovirus, it can occasionally be transmitted via seminal fluid and across the placenta. Overall, 24 serotypes have been reported around the world, generating only low levels of cross-protection and complicating vaccination strategies. Due to its economic impact, BT is an Office International des Epizooties (OIE)-listed disease. Economic losses associated with BTV infection are caused directly through reductions in productivity and death and more importantly indirectly through trade losses due to animal movement restrictions, restrictions on the export of cattle semen and the costs of implementing control measures, including diagnostic tests. The recent European outbreaks of BT started in 1998 with the arrival of BTV-9 in Greece. Since then, the epizootic situation regarding BTV in and around Europe has changed dramatically, with incursions of new serotypes into existing endemic zones and the emergence of the virus in previously disease-free areas, affecting immunologically naive herds and generating important economic losses.

Bluetongue virus (density 1.337 g/cm³) is a non-enveloped virus with a genome of approximately 19 200 base pairs composed of ten linear segments of double-stranded RNA (dsRNA), containing 57% AU and 43% GC, with conserved 5 and 3 terminal sequences (GUUAAA at 5 and ACUUAC at 3 ends of the positive strand). The 10 dsRNA segments are packaged within a triple layered icosahedral protein capsid (approximately 90 nm in diameter). The virus particle has been extensively studied at high resolution by cryo-electro-microscopy and X-ray crystallography techniques. The outer shell is composed of two structural proteins, 60 trimers of VP2 (111 kDa) and 120 trimers of VP5 (59 kDa). The intermediate layer consists of the major immunodominant VP7 structural protein (38 kDa), organized in 260 trimers forming a T = 13 icosahedral lattice covering the subcore. The subcore consists of the 12 decamers of the VP3 protein (100 kDa), one centered on each of the five fold axes of the icosahedral particle structure. The 120 molecules of VP3 can be considered as a 'pseudo T = 2' icosaedral lattice, which houses the viral genome segments and three minor proteins involved in transcription and replication, namely the RNA-dependent RNA polymerase (VP1, 149 kDa), the RNA capping enzyme (VP4, 76 kDa) and the dsRNA helicase (VP6, 36 kDa). VP7 forms the outer layer of the transcriptionally active virus 'core'. Non-structural proteins (NS1, NS2, NS3 and NS3A) probably participate in the control of BTV replication, maturation and export from the infected cell. Unlike most single stranded RNA (ssRNA) viruses, the orbiviruses are genetically and antigenically stable throughout infection; point mutations do not appear to arise in vivo, at least at the high frequency noted with many non-segmented ssRNA viruses.

BTV Structural Proteins

VP2

The trimers of the VP2 form 'triskelion' motifs (three interlocked spirals) on the outer layer. VP2 is responsible for receptor binding, hemagglutination and eliciting serotype-specific neutralizing antibodies. Recombinant VP2 has a strong affinity for glycophorin A, a sialoglycoprotein component of erythrocytes, an interaction that could be involved in BTV binding to erythrocytes. Furthermore, VP2 and glycophorin can inhibit BTV attachment to susceptible cells, suggesting that the BTV receptor involves VP2 interaction with a cell surface glycoprotein. Inside the cell, VP2 associates to vimentin, which allows the proper sub-cellular localization of the protein and the interaction of mature BTV particles to intermediate filaments. Disruption of the VP2/vimentin interaction by pharmacological inhibitors leads to blockade of the virus egress.

VP2 is the major determinant of BTV serotype, with a minor role for VP5. Phylogenetic comparisons of VP2 from the 24 reference strains show a perfect correlation between sequence variation in genome segment 2 (Seg-2), coding for VP2, and BTV serotype. Sequences of seg-2 from the 24 BTV types cluster as ten distinct evolutionary lineages, identified as nucleotypes A–J. The inter-serotype VP2 nucleotide sequences varied from 29% (BTV-8 and BTV-18) to 59% (BTV-16 and BTV-22). Sequencing and phylogenetic comparisons of VP2 also revealed significant variations between strains of the same serotype that were derived from different geographical areas, with a maximum of 30% nucleotide sequence variation within the same serotype. These geographical variations define eastern and western VP2 topotypes within individual serotypes. Oligonucleotide primers can be designed targeting Seg-2, that can be used in RT-PCR assays to facilitate typing of BTV field and vaccine isolates of each serotype and topotype. Despite the overall sequence variability, some features of VP2 appeared to be conserved across serotypes, including the hydrophobicity profile, charge distribution and the position of certain cysteine residues.

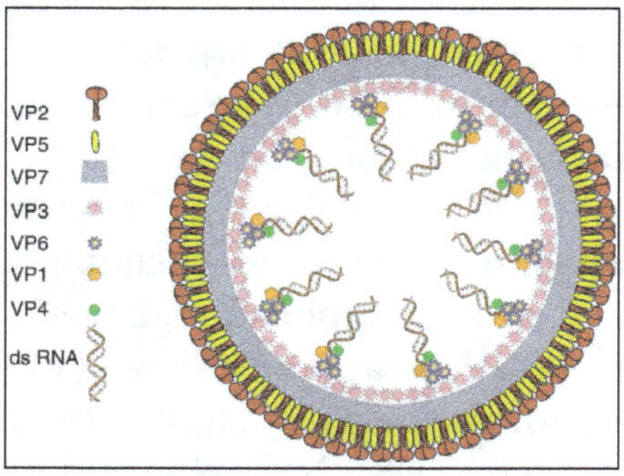

Representative scheme of BTV structural proteins and dsRNA segments.

VP5

In contrast to VP2, VP5 is significantly more conserved but shows some degree of variations that reflects the geographic origin. Trimers of VP5 form the globular motifs of the outer layer of the BTV virus particle. VP5 has recently been shown to be a membrane penetration protein that mediates release of viral particles from endosomal compartments into the cytoplasm. Analysis of the VP5 sequence using secondary structure prediction algorithms indicates that this protein is predominantly α-helical, with an amphipathic helical domain at the N terminus followed by a coiled coil domain, thus sharing structural features with class I fusion proteins of enveloped viruses. Furthermore, VP5 undergoes pH-dependent conformational changes that allow membrane fusion and syncytium formation. The syncytium formation by VP5 is inhibited in the presence of VP2 when expressed in a membrane-anchored form.

The Major Core Proteins VP3 and VP7

VP3 and to a lesser extent VP7 are conserved proteins and are hydrophobic in nature. They play an important role in the structural integrity of the virus core. They express group-specific antigenic determinants defining several distinct phylogenetic groups.

Importantly, cores are poorly infectious or even non-infectious in different mammalian cells but they are at least 100 fold more infectious for adult Culicoides midges or a Culicoides cell line (KC cells). VP7 can mediate attachment and penetration of insect cells in the absence of either VP2 or VP5, a process that may involve an arginine-glycine-aspartate (RGD) tripeptide motif present at amino acid residues 168 to 170 on the outermost surface of the VP7 trimers on the BTV core. VP7 can bind to glycosaminoglycans, although it appears likely that other specific receptors are also involved in cell attachment and penetration.

The VP3/VP7 complex protects the viral dsRNA genome from intracellular surveillance, thus preventing activation of type I interferon (IFN) production via cytoplasmic sensors such as cytosolic helicases, or interactions with dicer and RNA silencing mechanisms.

The Three Minor Core Proteins VP1, VP4 and VP6 (Transcription Complex)

VP1 is present in a low molar ratio (approximately 12 copies per particle) within the virion. VP1 can extend RNA synthesis from oligo(A) primers and acts as the BTV replicase that synthesizes dsRNA from a viral plus-strand RNA template. VP1 has an optimal activity at 27 °C to 37 °C, allowing efficient replication in both insect and mammalian cells.

The early BTV mRNA are capped. The cap (methylguanosine connected to the first nucleoside) stabilizes the mRNA and allows efficient translation. In cells, capping requires the action of four distinct enzymes. In BTV, all four reactions are catalyzed by the single VP4 protein, whose crystal structure shows an elongated modular architecture that provides a scaffold for an assemblage of active sites.

The VP6 protein has ATP binding activity and displays RNA-dependent ATPase and helicase functions. It unwinds duplexes of dsRNA and could assist mRNA synthesis from the genomic dsRNA template.

BTV Non-structural Proteins NS1, NS2, NS3 and NS3A

The two larger BTV non-structural proteins, NS1 and NS2, are the first and second most highly expressed proteins in infected cells, whereas the two closely related minor proteins NS3 and NS3A are barely detectable in mammalian cells. However, NS3 and NS3a are synthesised in much larger amounts in insect cells, suggesting that their role may be primarily related to BTV replication and dissemination within the insect vector.

NS1

Electron microscopic analysis of thin sections of BTV-infected cells have revealed a large number of virus-specific tubules (52.3 nm diameter and 1000 nm long) composed of multimers of the NS1 protein, a striking intracellular morphological feature of BTV infection. Expression of an scFv to NS1 reduced the cytopathic effects of BTV, suggesting that NS1 has a role in BTV cytopathogenesis. In addition, recombinant NS1 expressed in insect cells, which is highly immunogenic and easy to purify, has been used as a scaffold for exogenous peptides to generate heterologous recombinant protein vaccines against foot and mouth disease and influenza and HIV.

NS2

NS2 is the major constituent of the viral inclusion bodies (VIB) seen in infected cells mainly in the vicinity of the nucleus. NS2 binds to viral ssRNA and hydrolyses nucleotide triphosphates to nucleotide monophosphates. These two properties imply that NS2 might be involved in some way in selection and condensation of the BTV ssRNA segments prior to genome encapsidation. NS2 expression in cells is sufficient for formation of inclusion bodies and it recruits VP3, suggesting that NS2 is a key player in virus replication and core assembly.

NS3

NS3 and its shorter form, NS3A, which lacks the N-terminal 13 amino acids of NS3, are the only membrane proteins encoded by orbiviruses. Interestingly, NS3 and NS3A appear to be associated with smooth intracellular membranes, although they are also present at the plasma membrane. NS3 functions as a viroporin, facilitating virus release by inducing membrane permeabilization. In addition, NS3 binds to the cellular protein Tsg101, allowing BTV particles to also leave host cells by a budding mechanism, similarly to retroviruses. This budding mechanism might be involved in BTV egress from insect cells in which BTV does not induce significant cytopathic effect, whereas the viroporin mechanism would be more prominent in mammalian cells.

BTV Life Cycle

BTV interacts with the target cell surface via VP2 trimers binding to cell surface glycoproteins and possibly to other receptors. BTV core particles can also bind to cells (particularly insect cells) via VP7 trimers. The BTV particle is then internalized in endosomes via a clathrin-dependent endocytosis pathway. VP2 dissociates from the outer capsid layer in early endosomes. Acidification induces VP5 fusion with the endosomal membrane, delivering the transcriptionally active core into the cell cytoplasm. Like other members of the Reoviridae family, BTV replicates within the cytoplasm of infected cells. Within the BTV core, the VP1 molecules transcribe positive sense ssRNA copies from each of the ten BTV genome segments. These mRNA molecules are capped by the guanylyl-transferase and transmethylase activities of VP4 and leave the particles via channels situated at the five fold axes of the core particle. The viral mRNA serve as templates for translation in viral proteins, starting within two hours post infection. Viral positive RNA are directed to VIB where the correct encapsidation of the different segments (nature and numbers) within the VP3 shell may involve interactions with the helicase VP6, the ssRNA binding NS2 protein, and the VP1 and VP4 proteins. VP1 then synthesizes the negative strand RNA to produce dsRNA. It has been proposed that each dsRNA segment independently associates with a different transcription complex (VP1, VP4 and VP6) located at the inner side of VP3 along a fivefold axis, making a 'flower shape' in cryoelectromicroscopy.

Exchanges of dsRNA segments can occur when two different BTV (serotypes or strains) infect the same cells, contributing to the evolution of BTV through the process of reassortment. The process of reassortment may involve fusion of VIB formed with different viral particles. For unknown reasons, some segments are more often exchanged than others.

The VP3 subcores are relatively fragile and unstable structures that serve as a scaffold for the addition of VP7 trimers, giving rise to more rigid and stable cores. The outer capsid proteins VP2 and VP5 appear to be added to the progeny core particle surface at the periphery of the VIB as they enter the host cell cytoplasm. Mature progeny virus particles are transported within the cytoplasm on microtubules involving VP2/vimentin interactions. Release of virions from the infected cell occurs via cell membrane destabilization mediated by the NS3 viroporin activity, in some cases via budding, or as a result of cell death and lysis. Production of mature particles is exponential during the 8th and 24th hours post infection.

BTV Pathogenesis in Ruminants

Clinical Signs

In temperate climates in enzootic zones, outbreaks of BTV occur seasonally in association with high populations of the adult Culicoides vectors and can affect domestic and wild ruminants. Cattle, which are susceptible to BTV infection, usually do not develop overt clinical signs, but they can manifest an IgE-mediated hypersensitivity reaction. However, cattle are important in transmission, acting as reservoirs for the virus. The strain of BTV serotype 8 that has invaded northern Europe is unusual because a large number of infected cattle also developed clinical signs. Sheep and some wild ruminants display a variety of clinical manifestations, ranging from subclinical infections or mild disease, to acute or even fatal disease. Viraemia is usually detectable around 3–5 days post infection in sheep. Clinical signs include pyrexia, tachypnoea and lethargy. Gross pathological alterations are characterized by widespread oedema, haemorrhages especially in the lymph nodes, lungs, heart and skeletal muscles and necrosis of the mucosal surfaces in the oro-nasal and alimentary systems. Lungs, the shock organ in ruminants, are especially susceptible to permeability disorders of the vasculature induced by BTV. Microscopic lesions include endothelial hypertrophy, vascular stasis and thrombosis with tissue infarction. Animals that survive acute infection may develop chronic dermatitis, and vesicular and erosive lesions at interdigital and mucosal surfaces. A panleukopenia reaches its maximum at day seven to eight post infection and affects all lymphocytes, especially CD8 T cells.

The severity of the disease induced by BTV varies with host age and health status. Immunodeficiency due to concurrent infections can strongly exacerbate disease manifestations, as well as exposure to strong solar radiations.

Cell Tropism, Dissemination and Reservoir

After initial replication in the lymph nodes draining the sites of inoculation, BTV disseminates to secondary sites, principally the lungs and spleen, where it replicates in endothelium and mononuclear phagocytes. BTV can disseminate via lymph or/and via blood. BTV infects monocytes both in vivo and in vitro. In vivo, infectious BTV can be retrieved transiently (< 1 week) from monocytes (minimum 105 cells). Monocytes also express BTV antigens in vivo at low frequency (four NS2 antigen-positive monocytes per 2×10^5 peripheral blood mononuclear cells). In vitro, around 15% of monocytes express BTV VP7 after 36 h in culture and they produce low amounts of infectious BTV in vitro. Conversely, resting T lymphocytes are not efficient at supporting BTV replication unless they are activated by mitogens. Interestingly, T cell lines can be productively infected in vitro and blood T cells from infected sheep (3–13 days post infection) have been induced to produce infectious BTV when cocultivated with skin fibroblasts. However, it is unclear how monocytes and possibly blast T cells are involved in vivo in the pathogenesis of BTV. Last but not least, infectious BTV can also be detected in the intracellular vesicles of erythrocytes, in which it does not replicate but persists in invaginations of cell membrane. The association of infectious BTV with erythrocytes is detected very early after infection (24 h) and persists throughout viraemia.

Consequently, BTV infection in ruminants is characterized by a prolonged cell-associated viraemia that can persist in the presence of high titres of neutralizing antibody, although recovered animals are immune to re-infection with the homologous serotype of BTV. In sheep and cattle, infectious BTV can be detected in the blood for 35 to 60 days and viral structures for up to 160 days. It has been proposed that particles associated with erythrocytes are protected from early immune clearance. Furthermore, detection of BTV RNA up to 145 days after infection is remarkably similar to the lifespan of the ruminant erythrocytes, suggesting that erythrocytes are likely to be the critical mechanism that allows cattle to serve as natural reservoir hosts of BTV. Some other work indicated that BTV RNA can be detected even up to 222 days.

In enzootic areas, BT usually appears in late autumn after long periods of quiescence (8–12 months), a phenomenon called overwintering. However, conventional models for the transmission of BTV suggest that if adverse winter conditions last for more than 100 days, the virus should be unable to survive from one year to the next. Persistence of BTV in the larvae of vector Culicoides is considered to be highly improbable although it cannot be ruled out. Some authors have postulated that persistently infected $\gamma\delta$ T cells, recruited by midge biting, could facilitate the transmission of BTV for periods as long as nine weeks post infection in Dorset sheep. However, this finding was not confirmed in another study using a different sheep breed (Merino sheep).

Cellular Response to BTV

Viral haemorrhagic diseases are not only a consequence of direct viral damage to the cells but also the result of intense and sometimes deregulated inflammatory processes. BTV reduces endothelial electrical resistance in vitro associated to p38 MAP kinase-dependent cytoskeletal rearrangements. In addition, BTV infections induce cell death in many cell types and an important inflammatory cell response.

Cell Death

In mammalian cells, BTV induces cell death (apoptosis and necrosis) in cell lines, microvascular ovine and bovine endothelial cells, monocytes and in WC1-activated $\gamma\delta$ T cells. In mammalian cell lines, uncoating of BTV, but not BTV replication, is required to trigger apoptosis. Extracellular treatment with a combination of the viral outer capsid proteins VP2 and VP5, and the cell penetration protein VP5 is sufficient to trigger apoptosis, involving activation of NF-κB. However, some cell types that support the replication of the virus do not demonstrate a cytopathic effect to BTV, such as insect cells, T cell lines and activated blood lymphocytes. The budding versus the viroporin-mediated viral exit mechanism might partially explain this difference.

Cytokines and Prostanoid Induction

Infection of bovine and ovine microvascular endothelial cells induces the transcription of interleukin 1 (IL-1), IL-8, IL-6, cyclooxygenase-2, and inducible nitric oxide synthase. These mediators have been involved in the pathogenesis of severe viral haemorragic fevers. Infection of sheep and cattle with BTV induces a plasmatic increase of prostacyclin and thromboxane. Thromboxane is a strong pro-coagulant factor whereas prostacyclin is a potent vasodilatator and inhibitor of platelet aggregation. There is a much higher prostacyclin/thromboxane ratio in cattle, which may explain the lower sensitivity of cattle to BTV induced microvascular injury and thrombosis.

BTV is also a strong inducer of type I IFN in vivo, in sheep, cattle and mice. A strain of BTV serotype 8 was shown to be an extremely potent inducer in vivo in mice, even when UV-irradiated. Strains of many BTV serotypes have been reported as IFN inducers but they may differ in their ability to induce IFN depending on the cell context. A wide variety of cells produce type I IFN after BTV stimulation in vitro, including mouse embryo cells, human leucocytes, leucocytes from adult sheep. Double stranded RNA from BTV also induced IFN in mice. However, the exact viral component involved in the induction is unknown.

Immune Responses against BTV

Transfer of antibodies and T cells in monozygotic sheep have shown that both humoral and cellular immune effector mechanisms are able to protect sheep against BTV infection and disease.

Humoral Immunity against BTV

Passive serum transfer studies have shown that BTV specific antibodies can confer protection in a serotype specific manner, suggesting an in vivo role for antibody-mediated viral neutralization. However, exactly how antibodies interfere with BTV infection in vivo is unknown. Attempts to demonstrate the existence of antibody-dependent cell-mediated cytotoxicity (ADCC) and complement-facilitated antibody-dependent cell-mediated cytotoxicity in both cattle and sheep, using a variety of systems, have consistently failed.

VP2 and VP5 are the only BTV proteins shown to induce neutralizing antibodies with VP2 being the major protein involved in serotype specificity. Sheep inoculated with VP2, either isolated from intact particles or generated by in vitro expression, produced virusneutralizing antibodies and were resistant to challenge with the homologous BTV serotype. Two hydrophylic domains in VP2, AA 199–213 (region 1) and AA 321–346 (region 2), contain multiple conformation-dependent neutralization epitopes. Mutations in neutralization escape mutants clustered in these two regions, where the lowest level of AA identity between serotypes was also found. Interestingly mutations in the region 1 of the VP2 protein of the same original BTV serotype 17 generated two co-circulating strains that displayed distinct neutralization characteristics.

Co-expression of VP2 with VP5, or in combination with core proteins to form double-shelled virus-like particles enhanced the neutralizing antibody response of inoculated sheep as compared to VP2 alone. VP5 improved the neutralization of BTV, possibly through its conformational influence on VP2. In addition VP5 may directly induce neutralizing antibodies although neutralizing monoclonal antibodies to VP5 have not been described.

Neutralizing antibodies generally protect only against homologous virus, although serial infections of sheep with two serotypes can protect the animals against challenge with a third serotype. Depending on serotypes, the antibodies can protect against the infection by a limited number of other serotypes, associated to similarities of sequences in VP2. Since the level of neutralizing antibodies elicited by VP2 greatly varies between sheep, an outbred species, vaccine strategies would benefit from inclusion of both VP2 and VP5 antigens.

Cellular Immunity against BTV

The level of neutralizing antibodies does not always correlate with the degree of protection after vaccination with live BTV and some inactivated vaccine preparations can confer protection in the absence of detectable levels of neutralizing antibodies. Furthermore adoptive transfer of BTV-specific cytotoxic T lymphocytes (CTL) to monozygotic sheep has been shown to confer partial protection, indicating that CTL participate in protection.

Cytotoxic T cell Response in Mice and Sheep

Evidence that the cross-protective immune response to BTV involves CTL has been obtained both in laboratory mouse models and sheep.

In mice, statistical analysis of the CTL response indicated that non-structural protein peptides are the predominant source of homotypic and heterotypic CTL recognition, followed by the minor core inner capsid protein VP3 and finally the inner capsid protein VP7 and the outer capsid proteins VP2 and VP5. The outer capsid proteins VP2 and VP5 that vary considerably between serotypes were not recognized by heterotypic CTL. Furthermore, Baculovirus expressed NS2, NS1, NS3, VP1, V4 and VP6 induced a CTL response in mice, which was partially protective against a vaccinia virus expressing the homologous protein only in the case of NS2.

In sheep, BTV-specific cross-reactive CTL have been described and CTL lines have been shown to inhibit viral replication in skin fibroblasts infected with homologous and heterologous virus types. Recombinant vaccinia viruses expressing BTV antigens demonstrated that VP2 and NS1 are major CTL immunogens and that NS1-specific, but not VP2-specific CTL are cross-reactive. Furthermore, the non-structural protein NS1 was recognized by CTL from all Merino sheep immunized with recombinant vaccinia viruses, while VP2, VP3, VP5 and VP7 were recognized by CTL from only some sheep. The other proteins (NS2, NS3, VP4, VP6) did not appear to contain CTL epitopes, whereas NS2 induced CTL in mice. Thus, important variations in the protein targets of CTL exist between individuals and host species.

'Helper' T Cell Responses

Major 'helper' serotype-specific determinants are present in VP2 and some in VP5, while major serotype cross-reactive determinants are located within the core structural proteins. When sheep were vaccinated with a capripox virus encoding for VP7, clinical protection was obtained against heterotypic challenge, although the virus still replicated. The immune effectors involved in the VP7 induced protection were probably

CD4 T cells, but their direct functional contribution was not evaluated. Besides, VP7 is immunodominant during antibody production using in vitro priming techniques, but the antibodies are not neutralizing and probably not protective.

Vaccines against BTV

Much effort has been devoted to the development of vaccines against BT. The protective vaccines against BT in ruminants that are currently available commercially can be either attenuated or inactivated. Their protective activity is serotype-specific, probably due to the key role of the outer protein VP2 in the B and T cell mediated protective immunity. Thus, in endemic areas where multiple BTV serotypes may be present, efficient vaccines against several serotypes may be necessary. Furthermore, vaccines against BT need to be safe and should allow differentiating between vaccinated and infected animals (DIVA vaccines), in order to facilitate trade. Due to these three requirements (multi-serotypes, safety and DIVA properties), vaccination against BTV can be a difficult and controversial issue. Below, we summarize the findings of a recent review on the pros and cons of inactivated and attenuated vaccines.

Attenuated Vaccines

Attenuated vaccines produced by Onderstepoort Biological Products (South Africa) have long been used to control BT in sheep in southern Africa, and more recently in Corsica, the Balearic Islands and Italy. These vaccines provide robust protection after one injection for at least one year and they are cheap to produce. However, these vaccines are not always safe, especially in some susceptible breeds. Indeed their attenuation is difficult to control. They can generate mild clinical signs after injection, abortions, transiently depressed milk production and decreased semen quality. Moreover, as the vaccine virus can elicit a viraemia of over two weeks in vaccinated sheep, they can be spread to vectors with the potential for reversion to virulence and reassortment with genes of the wild-type virus giving rise to novel BTV strains with modified virulence. Finally, attenuated vaccines are not DIVA vaccines. For all these reasons, other vaccine strategies are required.

Inactivated Vaccines

Inactivated vaccines can generate a safe and protective immunity if properly prepared. However, a single vaccination may induce only the transient detection of neutralizing antibodies and it is usually insufficient to provide an immunity that lasts for several months (particularly in cattle). Nonetheless, a strong and long-term immunity can be elicited by two injections. DIVA inactivated BT vaccines are theoretically possible but have not yet been developed. Although inactivated vaccines are expensive to produce and their use is constraining, they are the best currently available safety/efficacy compromise. The European Food Safety Authority has recommended that countries use killed vaccines and they have been used since 2005 in some European countries (including France and Italy). However, the available inactivated vaccines are directed to only few serotypes.

Virus-like Particles

BTV structural proteins can be produced as recombinant proteins encoded by Baculoviruses in insect cells, in which they auto-assemble as virus like particles (VLP), presenting BTV antigenicity without BTV genetic information. They are thus considered to be naturally safe and they do not require any process of inactivation, although a recent study pointed out that laboratory-produced VLP batches included large quantities of Baculoviruses. Clinical batches should be carefully checked for the presence of Baculovirus and the potential risk of their replication in insect vectors in the field. Regardless, VLP are promising vaccine tools with which to generate multivalent BT vaccines, since VP2 from several viral strains can be included. VLP have been shown to be efficacious in protecting against homologous BTV challenge and partially protecting against heterologous BTV challenge in laboratory trials. Further studies are awaited to evaluate their structural long-term stability, their cost of production/purification and their efficacy in the field.

Recombinant Vectors

Recombinant vectors could be developed as future vaccines, if they are safe, inexpensive, DIVA, flexible for multi-serotype inclusions and if they provide long-term protective immunity in one shot. Although the road is still long to achieve these goals, some preliminary and promising laboratory studies have been published using poxvirus-derived vectors. Indeed, initial work showed that co-injection of vaccinia virus encoding for VP2 and VP5 proteins (three injections, 21 days apart) could confer protective immunity in sheep. More recently, a canarypoxbased vector that expressed optimized synthetic genes for VP2 and VP5 (two injections, 22 days apart) elicited high levels of neutralizing antibodies, a differential reactivity to VP7 as compared to sera from infected sheep (DIVA), and strong protection against homologous challenge (BTV-17); such a non replicative canarypox vector is extensively and safely used over the world in other recombinant vaccines. Finally, a replicative capripox encoding for VP2, VP7, NS1 and NS3 (one injection) was partially protective in sheep. Thus, recombinant vectors can provide protective immunity with DIVA properties but their efficacy barely reaches that of inactivated vaccines, still requiring several injections for efficient long-term protection. Recombinant vectors providing high levels of exogenous protein synthesis and generating a strong anamnestic cross protective immunity are awaited to be qualified for a reasonable development for field applications.

Culturing Viruses

Bacteriophage cultures require host cells in which the virus or phage multiply.

Strategies of Replication

Virus or phage cultures require host cells in which to multiply. For bacteriophages, cultures are grown by infecting bacterial cells. The phage can then be isolated from the resulting plaques in a lawn of bacteria on a plate.

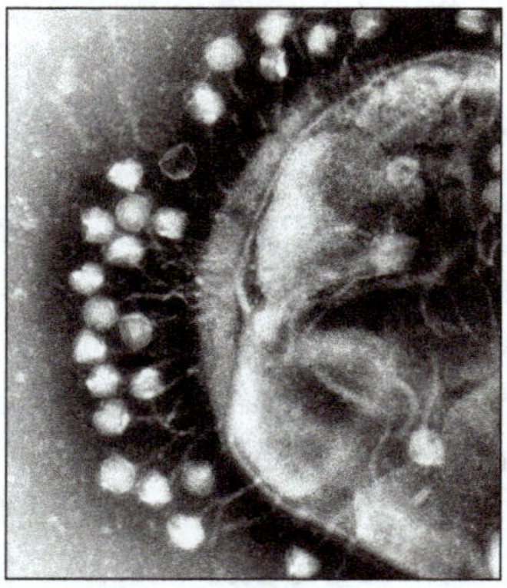

Bacteriophages infecting a bacteria: Virus or phage cultures require host cells in which to multiply. For bacteriophages, cultures are grown by infecting bacterial cells. The phage can then be isolated from the resulting plaques in a lawn of bacteria on a plate.

A bacteriophage is any one of a number of viruses that infect bacteria. They do this by injecting genetic material, which they carry enclosed in an outer protein capsid, into a host bacterial cell. The genetic material can be ssRNA, dsRNA, ssDNA, or dsDNA ('ss-' or 'ds-' prefix denotes single-strand or double-strand), along with either circular or linear arrangements.

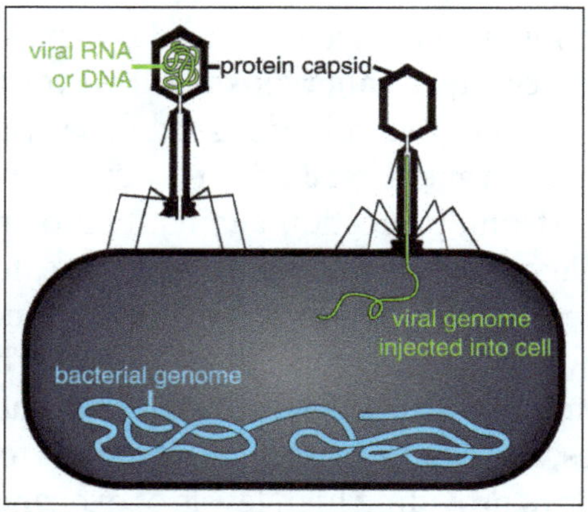

Bacteriophage: Diagram of how some bacteriophages infect bacterial cells.

To enter a host cell, bacteriophages attach to specific receptors on the surface of bacteria, including lipopolysaccharides, teichoic acids, proteins, or even flagella. This specificity means a bacteriophage can infect only those bacteria bearing receptors to which they can bind, which in turn determines the phage's host range. Host growth conditions also influence the ability of the phage to attach and invade them. As phage virions do not move independently, they must rely on random encounters with the right receptors when in solution within blood, lymphatic circulation, irrigation, soil water, or other environments.

Phages may be released via cell lysis, by extrusion, or, in a few cases, by budding. Lysis, by tailed phages, is achieved by an enzyme called endolysin, which attacks and breaks down the cell wall peptidoglycan. An altogether different phage type, the filamentous phages, make the host cell continually secrete new virus particles. Released virions are described as free, and, unless defective, are capable of infecting a new bacterium. Budding is associated with certain Mycoplasma phages. In contrast to virion release, phages displaying a lysogenic cycle do not kill the host but, rather, become long-term residents as prophage.

Tissue Culture of Animal Viruses

Viruses cannot be grown in standard microbiological broths or on agar plates, instead they have be to cultured inside suitable host cells.

Cell culture is the complex process by which cells are grown under controlled conditions, generally outside of their natural environment. In practice, the term "cell culture" now refers to the culturing of cells derived from multi-cellular eukaryotes, especially animal cells. However, there are also cultures of plants, fungi, and microbes, including viruses, bacteria, and protists. The historical development and methods of cell culture are closely interrelated to those of tissue culture and organ culture. Animal cell culture became a common laboratory technique in the mid-1900's, but the concept of maintaining live cell lines separated from their original tissue source was discovered in the 19th century.

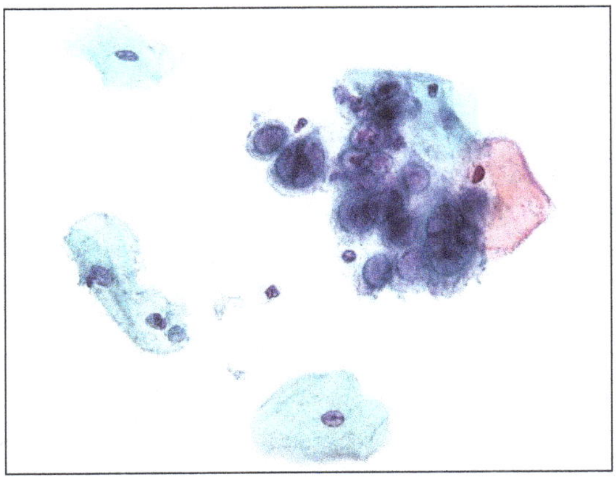

Cytopathic effect: Viral cytopathic effect of herpes simplex virus.

Viruses are obligate intracellular parasites that require living cells in order to replicate. Cultured cells, eggs, and laboratory animals may be used for virus isolation. Although embroyonated eggs and laboratory animals are very useful for the isolation of certain viruses, cell cultures are the sole system for virus isolation in most laboratories. The development of methods for cultivating animal cells has been essential to the progress of animal virology. To prepare cell cultures, tissue fragments are first dissociated, usually with the aid of trypsin or collagenase. The cell suspension is then placed in a flat-bottomed glass or plastic container (petri dish, a flask, a bottle, test tube) together with a suitable liquid medium. e.g. Eagle's, and an animal serum. After a variable lag, the cells will attach and spread on the bottom of the container and then start dividing, giving rise to a primary culture. Attachment to a solid support is essential for the growth of normal cells.

Cell cultures vary greatly in their susceptibility to different viruses. It is of utmost importance that the most sensitive cell cultures are used for a particular suspected virus. Specimens for cell culture should be transported to the laboratory as soon as possible upon being taken. Swabs should be put in a vial containing virus transport medium. Bodily fluids and tissues should be placed in a sterile container. Upon receipt, the specimen is inoculated into several different types of cell culture depending on the nature of the specimen and the clinical presentation. The maintenance media should be changed after one hour or the next morning. The inoculated tubes should be incubated at 35-37°C in a rotating drum. Rotation is optimal for the isolation of respiratory viruses and result in an earlier appearance of the cytopathic effects (CPE) for many viruses. If stationary tubes are used, it is critical that the culture tubes be positioned so that the cell monolayer is bathed in nutrient medium.

Inoculation of Live Animals

Live animal inoculation is a method used to cultivate viruses.

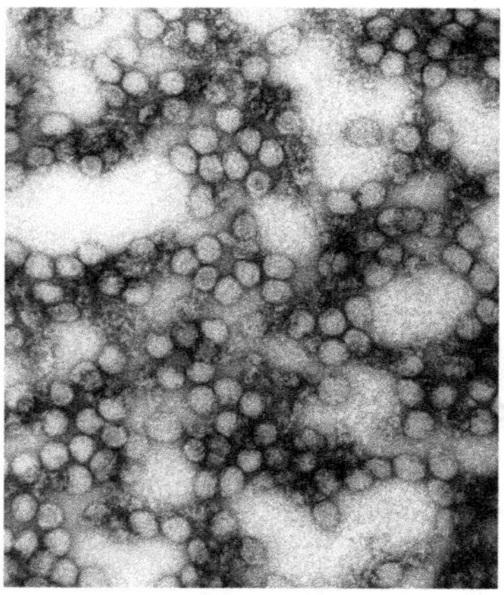

Yellow fever virus: A micrograph of the yellow fever virus.

Viruses are obligate intracellular parasites and cannot grow on inanimate media. They need living cells for replication, which can be provided by inoculation in live animals among other methods used to culture viruses (cell culture or inoculation of embryonated eggs). Inoculation of human volunteers was the only known method of cultivation of viruses and understanding viral disease. In 1900, Reed and his colleagues used human volunteers for their work on yellow fever. Due to serious risk involved, human volunteers are recruited only when no other method is available and the virus is relatively harmless. Smallpox was likely the first disease people tried to prevent by purposely inoculating themselves with other infections and was the first disease for which a vaccine was produced. Today, studying viruses via the inoculation of humans would require a stringent study of ethical practices by an institutional review board.

In the past few decades, animal inoculation has been employed for virus isolation. The laboratory animals used include monkeys, rabbits, guinea pigs, rats, hamsters, and mice. The choice of animals and route of inoculation (intracerebral, intraperitoneal, subcutaneous, intradermal, or intraocular) depends largely on the type of virus to be isolated. Handling of animals and inoculation into various routes requires special experience and training. In addition to virus isolation, animal inoculation can also be used to observe pathogenesis, immune response, epidemiology, and oncogenesis. Growth of the virus in inoculated animals may be indicated by visible lesions, disease, or death. Sometimes, serial passage into animals may be required to obtain visible evidence of viral growth. Animal inoculation has several disadvantages as immunity may interfere with viral growth, and the animal may harbor latent viruses.

Viral Identification

The genetic material within virus particles varies considerably between different types of viruses.

Replication of Viruses

The genetic material within virus particles and the method by which the material is replicated vary considerably between different types of viruses.

Types

DNA viruses: The genome replication of most DNA viruses takes place in the cell's nucleus. If the cell has the appropriate receptor on its surface, these viruses sometimes enter the cell by direct fusion with the cell membrane (e.g., herpesviruses) or, more usually, by receptor-mediated endocytosis. Most DNA viruses are entirely dependent on the host cell's DNA and RNA synthesizing machinery and RNA processing machinery; however, viruses with larger genomes may encode much of this machinery themselves. In eukaryotes the viral genome must cross the cell's nuclear membrane to access this machinery, while in bacteria it need only enter the cell.

RNA viruses: Replication usually takes place in the cytoplasm. RNA viruses can be placed into four different groups, depending on their modes of replication. The polarity of single-stranded RNA viruses largely determines the replicative mechanism, depending on whether or not it can be used directly by ribosomes to make proteins. The other major criterion is whether the genetic material is single-stranded or double-stranded. All RNA viruses use their own RNA replicase enzymes to create copies of their genomes.

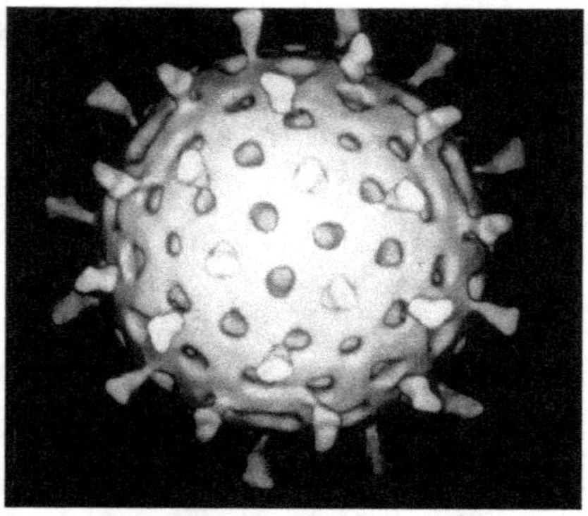

Computer assisted reconstruction of a rotavirus particle: An example of Baltimore Virus classification I: dsDNA virusesII; ssDNA virusesIII; dsRNA virusesIV; (+)ssRNA virusesV; (−)ssRNA virusesVI; ssRNA-RT virusesVII; dsDNA-RT viruses.

Reverse transcribing viruses: These have ssRNA (Retroviridae, Metaviridae, Pseudoviridae) or dsDNA (Caulimoviridae, and Hepadnaviridae) in their particles. Reverse transcribing viruses with RNA genomes (retroviruses), use a DNA intermediate to replicate, whereas those with DNA genomes (pararetroviruses) use an RNA intermediate during genome replication. Both types use a reverse transcriptase, or RNA-dependent DNA polymerase enzyme, to carry out the nucleic acid conversion. Retroviruses integrate the DNA produced by reverse transcription into the host genome as a provirus as a part of the replication process. Pararetroviruses do not, although integrated genome copies, usually of plant pararetroviruses, can give rise to infectious virus. They are susceptible to antiviral drugs that inhibit the reverse transcriptase enzyme, e.g. zidovudine and lamivudine. An example of the first type is HIV, which is a retrovirus. Examples of the second type are the Hepadnaviridae, which includes Hepatitis B virus.

The Baltimore classification developed by David Baltimore is a virus classification system that groups viruses into families, depending on their type of genome (DNA, RNA, single-stranded (ss), double-stranded (ds), etc.) and their method of replication. Classifying viruses according to their genome means that those in a given category will all behave in much the same way, which offers some indication of how to proceed with further research.

Virus Replication

Adsorption

Adsorption to the host cell surface is the first step in reproduction cycle of animal viruses. Adsorption of virion to the host cell surface takes place through a random collision of virion with a plasma membrane receptor site; the receptor is a protein, and frequently a glycoprotein. Animal viruses, like bacteriophages, possess attachment sides with the help of which it attaches to the receptor site.

Besides of glycoprotein receptors, sometimes, a complex carbohydrate (e.g., heparan sulfate) is the receptor, these receptors vary in their distribution pattern on plasma membrane and this distribution variation plays a key role in tissue and host specificity of animal viruses.

For instance, poliovirus receptors are found only in human nasopharynx, gut, and cells of spinal cord. While receptors of measles virus occur in most tissues.

Differences in nature of polio and measles can be explained through the dissimilarities in the distribution of receptor proteins of host cells to which viruses get adsorbed. In some naked viruses (e.g. adenoviruses) the attachment sites are small fibres at the corners of icosahedron. In enveloped viruses (e.g. myxoviruses) the attachment sites are the spikes present on the surface of envelope.

For example, influenza virus has two types of spikes: H (haemagglutinin) spikes and N (neuraminidase) spikes. The H spikes attach to the host cell receptor site by recognising sialic acid (N-acetyl neuraminic acid), the sugar derivative of glucoprotein.

Influenza neuraminidase helps the virus in penetrating the nasal and respiratory tract secretions by degrading mucosal polysaccharides. However, the receptor sites also vary from person to person.

Penetration

Animal viruses penetrate the host cell shortly after adsorption.

Though the detailed mechanism of penetration in not clear, the following three modes are the most favoured by the researchers:

- Direct penetration: Some naked animal viruses (e.g., picornaviruses, adenoviruses) use vesicle acidification that causes a major change in capsid structure after adsorption to plasma membrane. This altered capsid contacts the vesicle membrane and either releases the viral genome into the cytoplasm through a membrane pore (picornaviruses) or ruptures the membrane to release the viral genome (adenovirus) into the cytoplasm.

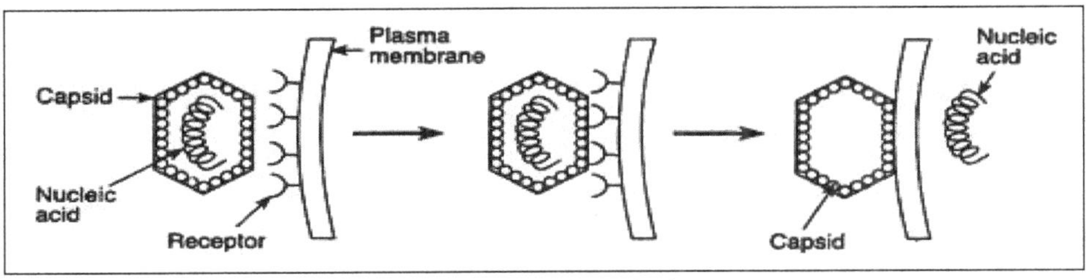

Direct penetration by naked virus.

- Fusion with plasma membrane: The envelop of enveloped virus (e.g., paramyxoviruses) fuses directly with host plasma membrane. Fusion may involve special envelop fusion proteins that bind to plasma membrane proteins. Finally, the nucleocapsid enters the cytoplasmic matrix where uncoating is done. A virus polymerase associated with the nucleocapsid, transcribes the virus RNA while the latter is still within the capsid.

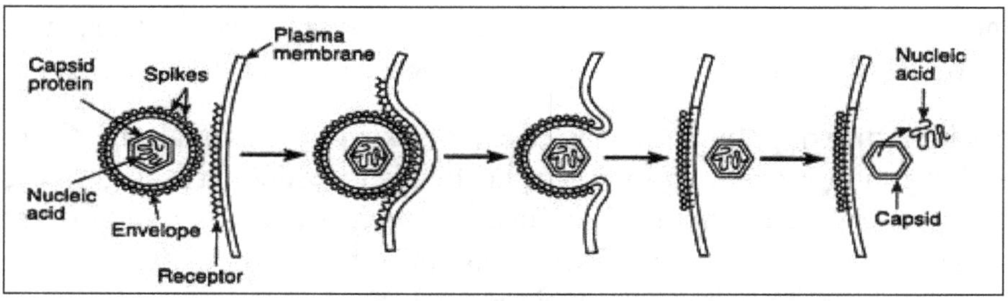

Entry of enveloped virus by fusion with plasma membrane.

- Endocytosis:Many of the enveloped viruses and certain non-enveloped viruses enter the host cell through engulfment by receptor-mediated endocytosis and form coated vesicles.

The virions attach to coated pits with the protein clathrin and the pits then pinch off to form coated vesicles filled with viruses. These vesicles fuse with lysosomes after the clathrin has been removed. Lysosomal enzymes help in uncoating of virion inside the cytoplasm.

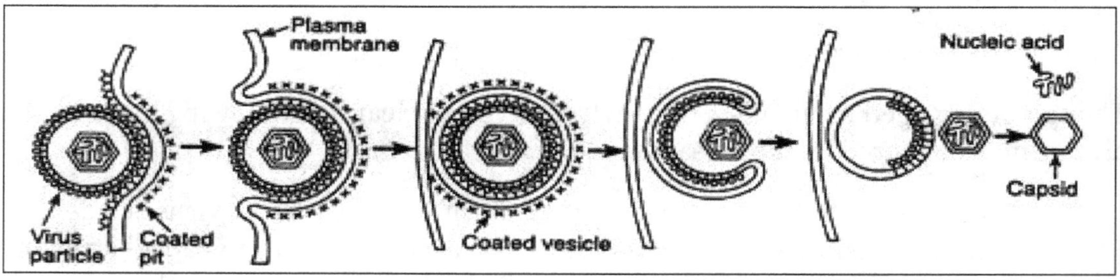

Entry of enveloped virus by endocytosis.

Uncoating

Uncoating is the process of separation of viral genome from the protein coat. Though the process of uncoating is not fully understood, it is proclaimed that the lysosomal enzymes help in animal virus uncoating by degrading the capsid and low endosomal pHs often trigger the process of uncoating.

It has been reported in some cases that the viral envelop fuses with the lysosomal membrane and the partially degraded capsid along with viral genome (nucleocapsid) is released into the host cytoplasm. Once in the cytoplasm, viral genome may be released from the capsid upon completion of uncoating or may function while still attacked to capsid components.

Replication of Viral Genome

The replication process of DNA viruses differs from that of RNA viruses. However, in some DNA viruses the replication takes place in cytoplasm (e.g., poxviruses) and in some others in the nucleus of host (e.g., parvoviruses, papovaviruses, adenoviruses, herpes viruses).

Replication of viral genome in RNA viruses is more or less the same as in DNA viruses except the mechanism of formation of mRNA among the different group.

Synthesis and Assembly of Virus Capsids

Certain late genes direct the synthesis of capsid proteins. The latter spontaneously self-assemble to form the capsid. It appears that in case of icosahedral viruses the capsid protein assembly first forms procapsid in which the viral genome is inserted by some unknown mechanism. However, in case of enveloped viruses the capsid protein assembly is generally similar to that of naked viruses (poxvirus is exception).

The capsids of these viruses are assembled in the cell cytoplasm by a lengthy, complex procedure that begins with the enclosure of a portion of cytoplasmic matrix through construction of a new membrane. Now the newly synthesized viral DNA condenses, passes through the membrane, and moves to the centre of the immature virus.

Release of New Virus

Release of newly formed animal viruses from the host cell differs between naked and enveloped viruses. The naked animal viruses are released most often by the lysis of the host cell. In enveloped viruses, however, the virus-encoded proteins are incorporated in the plasma membrane and then the nucleocapsid is simultaneously released; the envelope is formed by membrane-budding.

References

- Rhabdoviridae, ictv-reports-ictv-online-report-negative-sense-rna-viruses-mononegavirales: talk.ictvonline.org, Retrieved 21 June, 2019
- Footandmouth, diseaseinfo-243: thecattlesite.com, Retrieved 04 January, 2019
- Pestivirus, research-and-development-animal-health-welfare-and-biosecurity-diseases-reproductive: mla.com.au, Retrieved 25 February, 2019
- Arteriviridae, ictv-reports-ictv-9th-report-positive-sense-rna-viruses—posrna-viruses-220: talk.ictvonline.org, Retrieved 17 June, 2019
- Influenza-Infection-in-Animals, health: news-medical.net, Retrieved 16 May, 2019
- Culturing-viruses, boundless-microbiology: courses.lumenlearning.com, Retrieved 08 June, 2019
- Replication-of-animal-viruses-6-main-stages, viruses -54919: biologydiscussion.com, Retrieved 23 July, 2019

2

Zoonotic Viruses and Diseases

Zoonosis is a type of viral disease that spreads from animals to humans. It includes buffalo pox virus, yaba monkey tumor virus, mousepox, rabbitpox, swinpox virus, fibroma virus, pseudocowpox virus, asafarviradae, iridoviridae, etc. This chapter has been carefully written to provide an easy understanding of zoonotic viruses and diseases.

Buffalo Pox Virus

Buffalopox (BPX) is a contagious viral zoonotic disease primarily of domestic buffalo (*Bubalus bubalis*) characterized by disseminated pustular skin lesion mostly on the skin of udder, teat, medial aspect of thigh, inguinal region, base and inner surface of ear and eye. Disease occurs in two forms namely mild and severe. Lesions are localized in mild form whereas in severe form, which is uncommon, lesions are generalized. The secondary bacterial infection leads to concurrent otitis, conjunctivitis, mastitis as a complication of disease imposing significant economic losses in terms of reduced milk yield up to 50-80% in addition to reduction in working capacity of draft animal.

Transmission

Infection is carried from animal to animal through close contact with lesion, contaminated milker's hand, milking machine (teat cup) and mechanical transmission by biting flies.

Symptoms

Incubation period of the disease is 2-5 days. Typical pock lesions with raised edges on the skin of udder, teat, medial aspect of thigh, inguinal region, base and inner surface of ear and eye. The lesion passes through five stages during the course of the disease as follows:

- Stage I: Roseolar stage (diffuse redness on the skin).
- Stage II: Papular stage (elevated hard area on skin).
- Stage III: Vesicular stage (exudate enters in raised area).
- Stage IV: Pustular stage (pus formation).
- Stage V: Desquamative stage (crust or scab formation).

Course of the disease is two to three weeks.

Distribution

Since first reported in 1934 in Lahore (India), disease has been reported mainly sporadically and occasionally in epidemic form from various states of India including Haryana, Uttar Pradesh, Madhya Pradesh, Tamilnadu, Maharashtra, Karnataka, Andhra Pradesh, Gujrat and Rajsthan. Outbreaks have been recorded from many parts of the world including Egypt, Indonesia, Italy and Indian subcontinent.

Zoonotic Importance

BPX is an infectious Zoonotic disease primarily affecting the buffaloes and occasionally human and cows. In humans, disease is characterized by fever, axillary lymphadenopathy malaise and localized pock like lesions on hand, finger, face and legs. Incubation period in humans, 3-19 days, is comparatively longer than buffaloes.

Treatment and Control

BPX is a self-limiting disease with a course of two to three weeks but the secondary bacterial infection leads to productive losses in terms of mastitis.

- Segregation of affected and in contact animals from apparently healthy animals.
- Separate arrangement for milking of infected animals by separate attendants.
- Cleaning of pox lesions with potassium permanganate (1:1000) to inactivate the virus.
- Topical application of antiseptic ointment like boric acid (1:10) or broad spectrum antibiotic ointment on lesion to curtain secondary bacterial infection.
- Use of teat dips/antiseptic lotion.
- Treatment is symptomatic without any specific therapy.

Collection of Sample

Scab/crust, skin scrapings and vesicular fluid in sterile 50% buffered glycerol saline (pH7.4) with added antibacterial and antifungal antibiotics like penicillin, neomycin, polymixin and mycostatin. Keep sample on ice or frozen refrigerated packs in an insulated box for transportation to laboratory. Store the samples at -20 to -70 c for future use. Collect Paired sera samples (4ml) at a gap of 21 days, 1st at the onset of disease and 2nd after recovery from the disease.

Diagnosis

- Clinical sign: Typical pock like skin lesions.
- Laboratory confirmation of agent:
    - Virus isolation: Embryonated chicken egg inoculation through chorio- allantoic membrane (CAM) route or characteristic cytopathic effect (CPE) on Vero cell lines.
    - Electron microscopy: Brick shaped Virion.
    - Serological test to detect specific antigen and antibody against BPXV:
        - Counter immuno-electrophoresis (CIE).
        - Serum neutralization test (SNT).
        - Agar gel immuno diffusion (AGID).
        - ELISA.
    - Molecular techniques: Polymerase chain reaction (PCR).

Yaba Monkey Tumor Virus

The Yaba monkey tumor virus is a type of poxvirus. The first case of the virus was obtained from a colony of rhesus monkeys in Yaba, Lagos, Nigeria. The virus caused the formation of tumors on the bodies of the monkeys. From these tumors the virus was isolated and determined to be its own species of virus. It is a species of the Yatapoxvirus genus and is closely related to the tanapox. The virus gets its name from the suburb of Yaba, Lagos.

Signs and Symptoms

The Yaba monkey tumor virus is characterized by the formation of cutaneous histiocytomas. These cutaneous histiocytomas may form on the face, palms, digits, forearm, surfaces of the nose, sinuses, lips, palate. The tumors are visible within 4–5 days after infection of the virus. In rhesus monkeys the virus has a tendency to affect the face and sinuses causing large tumors to form. In general the only symptoms of the virus will be the formation of tumors.

Virus

The Yaba monkey tumor virus (YMTV) is considered a chordopoxvirus due to its ability to infect vertebrates and consists of linear double-stranded DNA. Being a part of the genus *Yatapoxvirus*, the YMTV is only able to infect primates - which includes humans.

The virus can be transmitted by direct contact on the skin or by transmission from mosquitos. The vaccinia virus is a known prevention method for the Monkey Poxvirus through intradermal inoculation, it has been found that intradermal inoculation of the vaccinia virus does not provide resistance to the Yaba monkey tumor virus however.

The tumors are known to resolve on their own after between 2–3 months after infection. Currently the best known treatment for the virus is the removal of the cutaneous tumors and subsequent treatment with antibiotics to prevent infection of the open wounds.

The best method of preventing infection is to wear gloves and protective clothing while handling infected primates.

The virus was first visualized after an outbreak of cutaneous histiocytomas in rhesus monkeys which were kept outdoors in 1956 Yaba, Lagos Nigeria. The virus has not been yet been observed in wild animals. This virus is transmittable to humans and has been transmitted in the past through handling of the monkeys, the symptoms are generally the same observed in the monkeys. There have only been a few cases of the virus transmitting to humans, with little documentation.

Mousepox

Ectromelia virus (ECTV) is a zoonotic viral disease. It belongs to the Poxviridae family of the genus Orthopoxvirus and is among the species of Vaccinia virus. Virions are oval or brick-shaped with a dimension of approximately 175 X 290 nm. It has a linear, double-stranded DNA genome that is 209,771 bp, surronded by a layer of lipids. This host-specialized virus infects with high efficiency, with the ability to spread systematically within its host and be effectively transmitted to others.

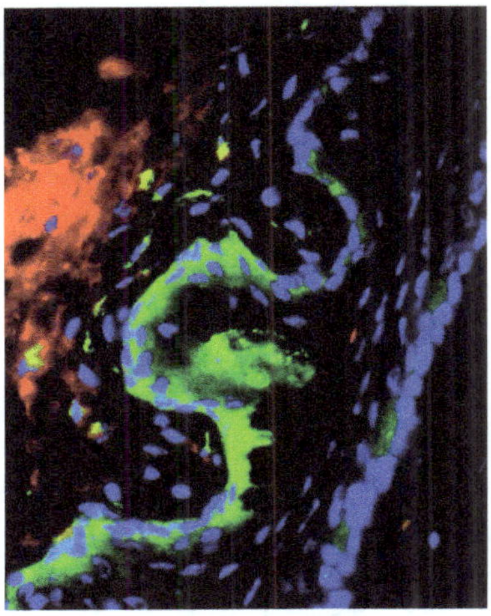

Ectromelia virus.

Characteristics of the Host

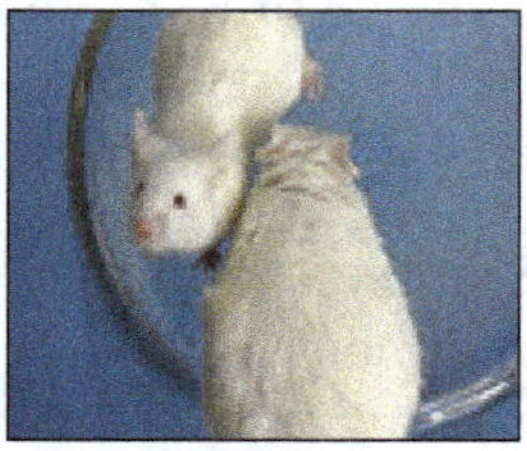

Mus musculus mouse CDC.

The host of ECTV is *Mus musculus*, better known as the mouse. While all mice are susceptible to the infection, clinical disease and mortality of mice is dependent upon virus and mouse strain. Mice strains highly susceptible to ECTV include A, CBA, C3H, DBA/2, and BALB/c while those that appear to be most resistant to infection include C57BL/6 and C57BL/10. The infection is not commonly seen among commercial colonies of mice but rather in research laboratories that exchange mouse tissues, live mice, transplantable mouse tumors, and mouse sere.

The natural reservoir of ECTV is unknown but it is suggested that wild mice may be involved as ECTV has a narrow host range infecting only certain species of mice. In laboratory studies, wild mice species including *Mus caroli*, *Mus cookii*, and *Mus cervicolor popaeus* are highly susceptible to experimental infection than other species of mice.

Host-symbiont Interaction

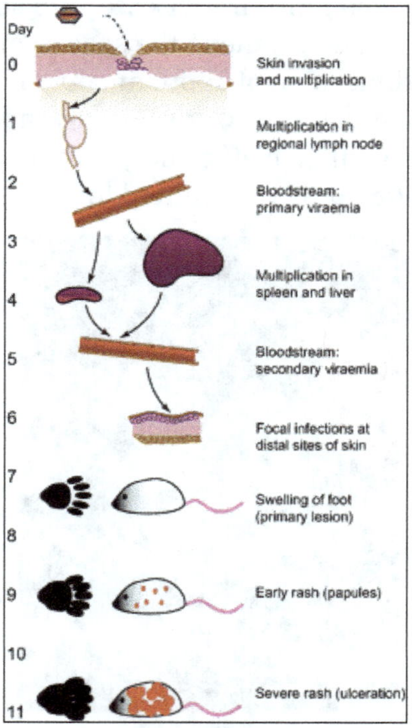

Transmission.

The interaction between the virus and the mouse is obligate pathogenesis. It can be transmitted through direct contact with an infected animal or through fomites. Transmission via the respiratory route has also been thought to be a possible route of entry as well. The skin is the natural route of the infection through abrasions in the skin. The virus is then able to replicate in the epidermis layer of the skin and then spread from the release of virual progeny from the initial infected site. Primary viraemia is a result of the virus' release into the bloodstream, causing infection of the spleen, liver, and other central organs. Secondary viraemia is due to the release of the virus from infected organs which results in infection of the skin. Clinical manifestations include swelling of the feet, amputation of the tail or feet, erosions, and encrustations on face, ears, feet or tail.

The incubation period for the virus is approximately 7 to 10 days. The infected animal will then begin to shed the virus and lesions that are characteristic of the virus begin to appear at the base of the tail. Swelling from the primary infection site will also occur as an inflammatory and immune response to the invading virus. Ultimately, infection leads to death. However, depending on the strain of ECTV mice can recover from the infection after three weeks to 116 days.

Molecular Insights into the Symbiosis

Specialized cell types in the epidermis layer of the skin of the mouse respond to infectious agents by expressing and then responding to signal molecules. Therefore, reactions to the infection occur because of the immune's response. One of the major epidermal cells of the mouse is keratinocytes. As micro abrasions occur in the skin it provides access for ECTV to access the lower levels of the skin. The abrasions uses the injury of keratinocytes to release stores of cytokines interleukin abbreviated IL. These possess pro-inflammatory responses which is dangerous to the virus in riding it of the host. Therefore, ECTV-encoding proteins inhibit IL and block the maturation of cytokines prior to the signalling in the infected cells.

Keratinocytes generates optimal gradient of chemokins to direct inflammatory cells and immune T cells to the initial infected site. ECTV encodes for a secreted CD30 homologue that is expressed by T cells as well as other cells types and is involved in antiviral immunity the induces reverse signaling.

Apoptosis induction is a cellular protective response that is used to eliminate virus-infected cells as well as limit virus replication in the skin. It can be induced through the binding of ligands to DED (death effector domain). Studies suggest that ECTV encodes for p28 gene which is suggested to have a role in the natural life cycle of ECTV help in inhibiting apoptosis.

What is the evolutionary history of the interaction? Do particular environmental factors play a role in regulating the symbiosis? The original strain of ECTV, the Hampstead strain, was first discovered in a laboratory-mouse colony in 1930 by Marchal in

England who called it "infectious ectromelia". Since its discovery, other ECTV strains have been isolated from various outbreaks around the world, with different disease severity. These include the Moscow, Hampstead and N1H79 strains, which are the most thoroughly studied and understood. Of these recognized strains, Moscow is the most virulent and infectious.

Within the orthopoxvirus family ECTV appears to be having the greatest phylogenetic distance compared to other species. However, there is little genetic diversity among the species of ECTV as they appear to be indistinguishable from the original Hampstead strain. Studies suggest that the simplest explanation for this lack of diversity is the fidelity of the poxvirus DNA polymerase.

Recent Discoveries

Moulton evaluated the ECTV inhibitor of complent enzymes (EMICE) that regulates complement activation on cells. This protects complement-sensitive intracellular virions from neturalization by serving as cofactors for the inactivation of essential pathways used to recover from infection.

Gratz evaluated how ECTV lives within the host. They confirmed that n1L, which is a virus ortholog in ECTV infection, is a virulence factor. It does this by interfering with the T well function of the host.

Rabbitpox

Rabbitpox virus is an Orthopoxvirus in the family *Poxviridae* and is antigenically related to vaccinia virus. The biological properties of the Utrecht and Rockefeller strains of rabbitpox viruses are indistinguishable from certain neurovaccinia strains. The close antigenic relationship between rabbitpox virus and vaccinia virus, taken together with the fact that all reported outbreaks of rabbitpox have occurred in laboratory colonies, suggests that rabbitpox may be a laboratory variant of vaccinia virus. Wittek used genome mapping to show that rabbitpox virus (Utrecht strain) was a strain of vaccinia virus. The viruses have been shown to exhibit over 95% sequence similarity.

Rabbitpox virus can be propagated on the chorioallantoic membrane of chicken embryos with development of distinct pocks.

A rare disease, rabbitpox has been seen in laboratory rabbits who have been infected by a virulent strain of the vaccinia virus or a virus that is closely related. Rabbitpox is a poxvirus (f. Poxviridae) and in the United States, a few outbreaks of this viral disease have been reported since 1930 when it was first diagnosed. Rabbitpox has also been seen in the Netherlands. Skin symptoms in the form of pox lesions may develop on the rabbit, and many also experience fever and nasal discharge (around two or three days after infection). The disease is very contagious and has a high rate of mortality. Though the disease is seen in laboratory rabbits, rabbitpox is not recognized in wild rabbits.

An acute generalized disease seen in laboratory rabbits, rabbitpox is caused by a virus closely related to vaccinia virus and includes skins lesions, fever and nasal discharge.

Symptoms of Rabbitpox in Rabbits

Rabbits that have rabbitpox will often develop skin lesions (around five days after first becoming infected) that are accompanied with fever, nasal discharge, facial swelling, lack of interest in eating and a lack of energy. Some rabbits may have pink or yellow lesions on their cornea and lesions elsewhere on their body may be obvious or only noticed when the rabbit is carefully examined. A rabbit may develop multiple pox lesions though that is not always the case; lesions may be in one place or throughout the rabbit's body. Usually there is a progression of papules to vesicles (smalls structures in the cell), which leads to crusting ulcers upon the abrupt separation of the vesicles. The significant swelling that a rabbit experiences as a result of rabbitpox will lead to his having enlarged lymph nodes.

- Lesions.
- Fever.
- Nasal discharge.
- Crusting.
- Enlarged lymph nodes.

For rabbits that don't survive the infection with rabbitpox, death will occur on average between seven to ten days after infection. Some rabbits lose their battle at around five days after infection and others several weeks after they have become infected.

Types

Rabbitpox can present with lesions or without lesions. Should there be no lesions present, the disease may be called pockless rabbitpox. Pockless rabbitpox occurred in a sudden outbreak in laboratory rabbits in Holland in 1941. Just like with rabbitpox, rabbits that developed pockless rabbitpox had a high mortality rate.

Causes of Rabbitpox in Rabbits

Rabbitpox is a poxvirus (*Poxviridae*) caused by a virus that is closely related to the vaccinia virus. Found in laboratory rabbits, the disease is known to be spread among rabbits through their grooming of one another as well as through their nasal discharge. The disease can also be spread by biting insects like mosquitoes. Rabbitpox has only been recorded in the United States and in the Netherlands and has only been seen in laboratory rabbits.

Diagnosis of Rabbitpox in Rabbits

A veterinarian will diagnose a rabbit with rabbitpox based on the symptoms he is exhibiting as well as through a histological examination. PCR assays may be available though they are not often obtained for clinical use. Shope fibromas and myxomatosis are other diseases that cause pox-like lesions, and these will have to be eliminated as possible diagnoses by the veterinarian prior to a diagnosis of rabbitpox being given.

Treatment of Rabbitpox in Rabbits

No treatment is typically provided for rabbits that are experiencing rabbitpox. Topical antimicrobial and anti-inflammatory ointments have been proven to be helpful for the rabbit in his recovery from the infection. Though it has been found to sometimes be poisonous to kidneys, cidofovir, an antiviral medication, has proven to be effective against poxviruses in rabbits, making it a possible treatment option. NSAID's, fluid therapy and assisted feeding are other options to help rabbits overcome the infection and feel more comfortable. Younger rabbits, and those that are pregnant or lactating, are the most likely rabbits to succumb to a rabbitpox infection.

Recovery of Rabbitpox in Rabbits

Rabbits infected with rabbitpox will benefit from supportive care. While no treatment is typically given, there are things that can be done to help the rabbit be more comfortable while he fights the infection. Should the rabbit need to be anesthetized while recovering, if his respiratory tract has been impacted by rabbitpox, the veterinarian will want to take special care. It is important to note that should a rabbit develop rabbitpox, once he has recovered he will not be a carrier of the infection.

The smallpox vaccine will be effective in rabbits for developing immunity to rabbitpox, making this an option that will eliminate the concern of rabbits contracting rabbitpox. Fortunately rabbitpox does not infect people so handling the rabbits during their infection with rabbitpox will not cause illness.

Sheep Pox and Goat Pox

Sheep pox virus (SPV) and goat pox virus (GPV) were once believed to be strains of the same virus, but genetic sequencing has now demonstrated them to be separate viruses. Most strains are host specific and cause severe clinical disease in either sheep or goats, while some strains have equal virulence in both species. Further complicating this is that recombination can occur between sheep and goat strains, which produce a spectrum with intermediate host preference and range of virulence.

SPV and GPV cannot be distinguished from each other with serological techniques, including viral neutralisation. SPV and GPV are also closely related to lumpy skin disease virus in cattle (LSDV), but there is no evidence LSDV causes disease in sheep and goats. It has a different transmission mechanism (insects) and partially different geographic distribution.

Resistance to Physical and Chemical Action

Temperature:	Susceptible to 56 °C/2 hours; 65 °C/30 minutes. Some isolates inactivated at 56 °C/60 minutes.
pH:	Susceptible to highly alkaline or acid pH (hydrochloric or sulphuric acid at 2% for 15 minutes).
Disinfectants/ chemicals:	Inactivated by phenol (2%) in 15 minutes. Sensitive to detergents, e.g. sodium dodecyl sulphate. Sensitive to ether (20%), chloroform, formalin (1%), and sodium hypochlorite (2-3%), iodine compounds (1:33 dilution), Virkon® 2%, quarternary ammonium compounds 0.5%.
Survival:	Susceptible to sunlight, but remains viable in wool/hair and dry scabs on skin for up to 3 months. Persists in unclean shaded pens for as long as 6 months. Survives freeze–thaw cycles, but infectivity may be reduced.

- Morbidity rate: Endemic areas 70–90%.
- Mortality rate: Endemic areas 5–10%, although can approach 100% in imported animals.

Hosts

- All breeds of domestic and wild sheep and goats, although most strains cause more severe clinical disease in only one species.
- Native breeds in endemic areas are far less susceptible than introduced breeds of European or Australian origin – morbidity and mortality may approach 100%.

Transmission

- Transmission is usually by aerosol after close contact with severely affected animals containing ulcerated papules on the mucous membranes. There is no transmission in the prepapular stage, e.g. animals early in disease or those dying peracutely (e.g. Soay breed of European sheep). There is reduced transmission once papules have become necrotic and neutralising antibody produced (about one week after onset). Animals with mild localised infections also rarely transmit disease.

- Infection may also occur through other mucous membranes or abraded skin.
- Chronically infected carriers do not occur.
- Indirect transmission by contaminated implements, vehicles or products (litter, fodder) occurs.
- Indirect transmission by insects (mechanical vectors) has been established (minor role).

Sources of Virus

- Ulcerated papules on mucous membranes prior to necrosis.
- Skin lesions with scabs: contain large amounts of virus in association with antibody but infectivity is not known.
- Saliva, nasal and ocular secretions.
- Milk, urine, faeces.
- Semen or embryos: Transmission not yet established.

Occurrence

Capripox is endemic in Africa north of the Equator, the Middle East, Turkey, Iran, Iraq, Afghanistan, Pakistan, India, Nepal, parts of the People's Republic of China, Bangladesh.The most recent outbreaks occurred in Vietnam in 2005, Mongolia in 2008 and 2009, and Azerbaijan in 2009. The first outbreak in Chinese Taipei occurred in 2008 and was eradicated by stamping out and movement control.

Diagnosis

Incubation period is 8–13 days. It may be as short as 4 days following experimental infection by intradermal inoculation or mechanical transmission by insects.

Clinical Diagnosis

Clinical signs vary from mild to severe, depending on host factors (e.g. age, breed, immunity) and viral factors (e.g. species predilection and virulence of viral strain). Inapparent infections also occur.

- Early clinical signs:
    - Rise in rectal temperature to above 40 °C.
    - Macules develop in 2-5 days: Small circumscribed areas of hyperaemia, most obvious on unpigmented skin.

    - Papules develop from macules: Hard swellings of between 0.5 and 1 cm in diameter – which may cover the body or be restricted to the groin, axilla and perineum. Papules may be covered by fluid-filled vesicles, but this is rare. A flat haemorrhagic form of capripox has been observed in some breeds of European goat, in which all the papules appear to coalesce over the body; this form is always fatal.

- Acute phase: Within 24 hours after appearance of generalised papules.
    - Affected animals develop rhinitis, conjunctivitis and enlargement of all superficial lymph nodes, especially prescapular lymph nodes.
    - Papules on the eyelids cause blepharitis of varying severity.
    - Papules on the mucous membranes of the eyes and nose ulcerate, creating mucopurulent discharge.
    - Mucosae of the mouth, anus, and prepuce or vagina become necrotic.
    - Breathing may become laboured and noisy due to pressure on the upper respiratory tract from the swollen retropharyngeal lymph nodes draining developing lung lesions.
- If animal survives acute phase:
    - Papules become necrotic from vascular thrombosis and ischaemic necrosis.
    - Papules form scabs in the next 5–10 days, which persist for up to 6 weeks, leaving small scars.
    - Skin lesions are susceptible to fly strike.
    - Secondary pneumonia is common.
    - Anorexia is unusual unless mouth lesions physically interfere with feeding.
    - Abortion is rare.
- Skin lesions: Congestion, haemorrhage, oedema, vasculitis and necrosis. All the layers of epidermis, dermis and sometimes musculature are involved.
- Lymph nodes draining infected areas: Enlargement (up to eight times normal size), lymphoid proliferation, oedema, congestion, haemorrhage.
- Pox lesions: On mucous membranes of the eyes, mouth, nose, pharynx, epiglottis, trachea, on the rumenal and abomasal mucosae, and on the muzzle, nares, in the vulva, prepuce, testicles, udder, and teats. Lesions may coalesce in severe cases.

- Lung lesions: Severe and extensive pox lesions, focal and uniformly distributed throughout the lungs; congestion, oedema, focal areas of proliferation with necrosis, lobular atelectasis. Enlargement, congestion, oedema and haemorrhages of mediastinal lymph nodes.

Differential Diagnosis

The clinical signs of severe sheep pox and goat pox are highly characteristic. However, in their mild form they can be confused with parapoxvirus causing orf or urticaria from multiple insect bites.

- Contagious ecthyma (contagious pustular dermatitis or orf).
- Insect bites.
- Bluetongue.
- Peste des petits ruminants.
- Photosensitisation.
- Dermatophilosis.
- Parasitic pneumonia.
- Caseous lymphadenitis.
- Mange.

Laboratory Diagnosis

Samples

Samples for virus isolation must be sent to the laboratory as soon as possible. They should be kept cold and shipped on gel packs. If these samples must be shipped long distances without refrigeration, glycerol (10%) can be added; tissue samples must be large enough that glycerol does not penetrate into the centre of the tissue and destroy the virus.

Neutralising antibodies can interfere with virus isolation and some antigen-detection tests; samples for these tests must be collected during the first week of illness. Samples for PCR can be taken after neutralising antibodies have developed. Paired serum samples should be collected for serology.

- Live animals: Full skin thickness biopsies, vesicular fluid if available, scabs, skin scrapings, lymph node aspirates, whole blood collected into heparin or EDTA, paired sera.
- Animals at necropsy: Skin lesions, lymph nodes, lung lesions, histology: Full set of tissues, especially those with lesions.

Procedures

Identification of the agent:

- Genome detection by polymerase chain reaction (PCR) using capripoxvirus-specific primers for the attachment protein gene is described in lumpy skin disease (chapter 2.4.14) of the OIE Terrestrial Manual.
- Transmission electron microscopy: Rapidly identifies typical capripox virions.
- Virus isolation in cell culture (primary lamb testis or lamb kidney): the appearance of CPE may take 4–12 days, intracytoplasmic inclusions are clearly seen by haematoxylin and eosin staining, and antigen can be detected by immunoperoxidase or immunofluorescence staining techniques.
- Agar gel immunodiffusion test (AGID): Tests lymph gland biopsy material taken from an early case of capripox; cross-reacts with parapox.
- Capripoxvirus antigen and inclusion bodies may also be seen in stained cryostat or paraffin sections of biopsy or post-mortem lesion material.
- Inhibition of cytopathic effect using positive serum.
- Antigen-detection ELISA.

Serological tests:

- Virus neutralisation: Most specific serological test, but not sufficiently sensitive since immunity to capripox infection is predominantly cell mediated – infected animals may only produce undetectable low levels of neutralising antibody.
- Indirect fluorescent antibody test: Cross reacts with other poxviruses.
- Agar gel immunodiffusion (AGID): Cross reacts with other poxviruses.
- Western blotting: Uses P32 antigen of capripoxvirus for reaction with test sera; sensitive and specific, but is expensive and difficult to carry out.
- ELISA: P32 antigen or another appropriate antigen expressed by a suitable vector could be used to develop an acceptable and standardised serological test.

Prevention and Control

Sanitary Prophylaxis

- If culling is not possible, isolation of infected herds and sick animals for at least 45 days after recovery.
- Slaughtering of infected herd if possible.

- Proper disposal of cadavers and products – burning or burial is often used.
- Stringent cleaning and disinfection of farms and equipment.
- Quarantine of new animals before introduction into herds.
- Animal and vehicle movement controls within infected areas.
- Vaccination may be considered when the disease has spread more widely.

Medical Prophylaxis

Live and inactivated vaccines have been used for the control of capripox. All strains of capripoxvirus so far examined share a major neutralisation site and will cross protect.

- There are several attenuated virus vaccines delivered by subcutaneous or intradermal route; conferred immunity lasts up to 2 years.
- Inactivated vaccines give, at best, only short-term immunity.
- Currently, no recombinant vaccines for capripoxviruses are commercially available. However, a new generation of capripox vaccines is being developed that uses the capripoxvirus genome as a vector for the genes of other ruminant pathogens, for instance genes of rinderpest and peste des petits ruminants (PPR) viruses.

Swine Pox Virus

A viral disease that is characterized by the acute appearance of round to oval cutaneous lesions that heal in three to four weeks.

Occurrence

Swine pox occurs only in swine. All age groups are susceptible but most outbreaks are encountered in young growing pigs. It occasionally occurs congenitally and in neonates. Swine pox occurs worldwide in the major swine-raising countries. It occurs more often where insect vectors of pox virus are not controlled. Modern swine raised free of lice rarely exhibit the disease.

Swine pox was first reported in Europe in 1842 and in the United States in 1929. The disease persisted and was relatively common, perhaps because it caused little obvious economic loss to swine producers. Outbreaks once were caused by two viruses: vaccinia virus, a laboratory strain, and swine pox virus. Vaccinia virus is no longer used to vaccinate people against small pox; hence it no longer spreads to pigs. Only swine pox virus now causes pox in swine.

Swine pox virus, a member of the Poxviridae family, is the etiologic agent. The virus is rather resistant to environmental degradation; it persists in scabs for up to a year. Microscopically, the virus is visible as large intracytoplasmic inclusions in infected cells.

Epidemiology

Persistent virus within desquamated scabs can be transmitted to other pigs when introduced into skin abrasions. The virus can also spread horizontally from nasal and oral secretions from infected pigs. Once a few virus-containing lesions appear, the virus can be transmitted mechanically by the bites of vectors that have recently fed in pox lesion areas. Common vectors include the hog louse (*Haematopinus suis*), mosquitoes, and biting flies. Once introduced, infected swine or virus-carrying biting insects can initiate outbreaks in susceptible herds. There is evidence of transplacental infection of neonatal pigs.

Pathogenesis

Virus introduced into abrasions replicates in cells of the stratum spinosum causing typical skin lesions to develop. The lesions go through the stages of macules, papules, vesicles, pustules and crusts. All but the vesicle stage are easily identified. As lesions mature and heal, crusts containing virus desquamate and leave white macules.

A viremic stage of the disease has not been demonstrated. However, it must occur in some animals since transplacental infection of neonatal piglets has been reported. Antibody to pox virus is formed as a consequence of infection and immunity is long lasting.

Clinical Signs and Lesions

Systemic clinical signs of disease seldom are observed but the skin lesions are easily seen. Initial lesions usually are along the ventral aspect of the abdomen, inside the legs and in inguinal areas, all favorite feeding sites of hog lice. However, lesions have been observed at many cutaneous sites so most of the skin appears susceptible. Initially, there are small vesicles that largely are unnoticed. The papules that result are seen as 1-3 mm circular red spots. As lesions mature, crusts form that eventually heal uneventfully. In severe infections, lesions may occur in upper respiratory and digestive tracts. This is particularly true of congenitally-acquired pox where piglets are born with or develop lesions over the entire body shortly after birth. Exudative epidermitis (greasy pig disease) and secondary bacterial dermatitis occasionally occur as a sequel to swine pox. Swine pox does not appear to cause a pruritic response in affected pigs.

Diagnosis

Swine pox is easily diagnosed by identifying the lesions. Lesions are round to oval, usually less than 1.0 cm in diameter. The brown to black crusts are easily seen but the vesicle stage seldom can be identified grossly. If confirmation is needed, histologic diagnosis can be made by identifying typical, large, intracytoplasmic inclusion bodies in cutaneous lesions. Pox lesions should be differentiated from those of other vesiculating viral diseases, pityriasis rosea, dermatosis vegetans, streptococcal dermatitis, ringworm and greasy pig disease.

Control

Control of swine pox relies on herd immunity and prevention of transmission. Pox seldom is a problem if the usual vectors of virus, especially hog lice, are controlled by insecticides or elimination. Congenital swine pox is generally a sporadic, self-limiting event. There are no treatments.

Fibroma Virus

Discovered in 1931 by Shope, this fibroma virus regularly affects wild rabbits and occasionally domestic ones too. While these fibromas are generally harmless, there are cases where they can lead to more serious issues. In young rabbits, the virus can progress into general and severe disease, including cancerous spreading tumors that can lead to death. In adults, tumors can progress into malignant tumors, or they can grow too large, invading the surrounding tissues and causing damage to muscles. Removal of fibromas can be effective in some cases.

The Shope fibroma virus in rabbits is often confused with the oral papillomavirus or the Shope papillomavirus, also called cottontail rabbit papilloma. However, the Shope fibroma virus is a distinct virus which stems from the pox virus family, while papillomas are papovaviruses.

The Shope fibroma virus that infects rabbits is a pox family virus, in the leporipoxvirus group. Once a rabbit is infected, fibrous tumors called fibromas begin to form in the skin. Often appearing on the legs, feet, ears and face of adult rabbits, these benign tumors naturally regress after a period of months, leaving the rabbit clinically normal. Affected rabbits show no other symptoms, and generally treatment is not given.

Symptoms of Shope Fibroma Virus in Rabbits

- Thickening of skin.
- Swelling of skin.
- Skin tumors on legs, feet, muzzle, ears, face, and eyelids.
- Tumors that are firm and flattened, pinkish red, ovoid or spherical, and are covered with a thick crust.

- Hair loss near tumors.
- Skin ulceration near tumors.

Young rabbits that are affected by this virus can have more severe symptoms that include:

- Generalized disease.
- Tumors in organs.
- Inflamed and degenerative lesions resembling myxomatosis.
- Malignant tumors that are widespread.
- Death.

Causes of Shope Fibroma Virus in Rabbits

Discovered in 1931 by Shope, this fibroma virus regularly affects wild rabbits, and occasionally domestic ones too. While these fibromas are generally harmless, there are cases where they can lead to more serious issues. In young rabbits, the virus can progress into general and severe disease, including cancerous spreading tumors that can lead to death. In adults, tumors can progress into malignant tumors, or they can grow too large, invading the surrounding tissues and causing damage to muscles. Removal of fibromas can be effective in some cases.

The Shope fibroma virus in rabbits is often confused with the oral papillomavirus or the Shope papillomavirus, also called cottontail rabbit papilloma. However, the Shope fibroma virus is a distinct virus which stems from the pox virus family, while papillomas are papovaviruses.

Diagnosis of Shope Fibroma Virus in Rabbits

Diagnosis of the Shope fibroma virus in your rabbit begins with the presence of a fibroma, and any other accompanying symptoms. The fibroma is examined, and often a biopsy sample is taken for testing. It is important to differentiate this condition from myxomatosis and papillomatosis. This is done by assessing the characteristics of the fibroma, virus isolation, serologic tests that use a fibroma antibody, and a PCR test.

If the fibroma has progressed to a malignant state, other tests may be performed to assess if it has spread to other parts of the body. Such tests can include X-rays, ultrasounds, CT scans, and MRIs.

The fibromas usually regress completely after 6 to 14 months, leaving the affected rabbit healthy, and immune to further infections of the same virus. Because of this, and the fact that affected rabbits generally have no other symptoms, the Shope fibroma virus is not treated. Since this condition allows complete self-recovery, there are currently no control measures in effect.

However, there are cases that may require surgical removal. If the fibroma has grown to a large enough size that it disturbs your rabbit's daily activities, or if it begins to compress, damage or invade muscles and tendons, removal may be recommended. If the fibroma has become malignant, or has spread in the body, treatment will follow the standard protocol for a condition of cancer. This includes surgical removal of the growth, and supplementary care as needed.

An associated virus, malignant rabbit fibroma virus, is believed to be the result of a combination of the fibroma virus and the myxoma virus, and is currently under study.

Recovery of Shope Fibroma Virus in Rabbits

The recovery of the benign fibromas that result from most cases of the Shope fibroma virus in rabbits is good, as the fibromas recede on their own over a period of several months. If a malignancy was found and removed, recovery is dependent on how far the tumors have spread. Your veterinarian will discuss your rabbit's recovery, and any care your rabbit may need once you return home. This can range from antibiotics, to changing dressings, to supplementary care.

Prevent your rabbit from contracting the Shope fibroma virus by keeping him away from tick and mosquito ridden areas, or use insect control.

Squirrel Fibroma Virus

Squirrel pox, or squirrel fibroma is a viral disease which produces multiple tumors on the skin of gray and fox squirrels. It belongs to the pox group of viruses which produce rabbit fibromatosis, rabbit myxomatosis, hare fibromatosis and some deer fibromas. Multiple skin tumors on gray squirrels from Maryland were reported in 1953. Since then, reports of squirrels with skin tumors have come from Florida, New York, Virginia, North Carolina, and Ontario. In Michigan, they have been found on both gray and fox squirrels in the Lower Peninsula.

Transmission and Development

These tumors are presumably all caused by a virus. Although the natural history of the virus is not known, both Aedes aegypti and Anopheles quadrimaculatus mosquitoes have transmitted it from squirrel to squirrel in the laboratory. It is probably transmitted by insects in nature as well. The tumors develop at the site where the insects feed on the animal.

Clinical Signs and Pathology

Naturally infected squirrels are mostly juveniles. The tumors may be scattered over all the body and range in size from a few mm to 25mm in diameter. Metastasis to the lungs, liver, kidney, and lymph nodes has been reported, but rarely. In general, there are no obvious signs of illness in naturally infected squirrels except for the presence of the tumors over the skin. In severe cases, when vision is obstructed or the skin becomes secondarily infected, the animal may be less active, weak, and eventually die. The microscopic lesions of squirrel pox virus in the gray squirrel are similar to those reported from Shope's rabbit fibroma virus in the cottontail.

Diagnosis

The disease is diagnosed by finding characteristic gross and microscopic lesions. Confirmation of the disease is either by histologic examination of tissues for intracytoplasmic viral inclusion bodies or by virus isolation.

Treatment and Control

There is no known treatment; presumably a vaccine could be developed. However, it would not be logistically or economically feasible to treat free-living wild animals. In cases where the disease is not severe, the lesions probably regress and the animal recovers completely.

Significance

As far as we know, the virus only infects squirrels in nature. However, in the laboratory it has been successfully transmitted to woodchucks and rabbits.

In Michigan, where squirrel pox is common, there apparently have been no noticeable effects on squirrel populations. The disease is of no public health significance. The carcasses of affected animals are safe for human consumption, since current knowledge indicates the virus is not transmissible to man. Furthermore, the tumors are usually confined to the skin and are removed when the animal is skinned. If metastasized areas are found in the viscera, even though the lesions are not of public health significance, for aesthetic reasons the carcass should be discarded.

Fowlpox

The morphology of the fowlpox virus is like that of other viruses of the Poxviridae family. The mature virus (elementary body) is brick shaped and measures about 330 × 280 × 200 nm. The outer coat is composed of random arrangements of surface tubules. The virion consists of an electron-dense centrally located biconcave core or nucleoid with two lateral bodies in each concavity and surrounded by an envelope. The 288 kbp fowlpox virus genome encodes for over 250 genes.

Fowlpox has a world-wide distribution and is caused by a DNA virus of the genus Avipoxvirus of the family Poxviridae. Its incidence is variable in different areas because of differences in climate, management and hygiene or the practice of regular vaccination. It can cause drops in egg production, or retarded growth in younger birds.

Fowlpox is a slow-spreading virus disease of chickens and turkeys, characterised in the cutaneous form (dry pox) by the development of proliferative lesions, ranging from small nodules to spherical wart-like masses on the skin of the comb, wattle and other unfeathered areas. In the diphtheritic form (wet pox), slightly elevated white opaque nodules develop on the mucous membranes. They rapidly increase in size to become a yellowish diphtheritic membrane. Lesions occur on the mucous membranes of the mouth, oesophagus, larynx or trachea. The mortality rate is higher in the diphtheritic form than in the cutaneous form, sometimes nearing 50% particularly in young birds. Integration of reticuloendotheliosis virus (REV) sequences has been observed in the genome of fowlpox virus. It is interesting that this insertion event occurred over 50 years ago. While most field strains of fowlpox virus contain REV provirus, vaccine strains have only remnants of long terminal repeats. Virulence is enhanced by the presence of REV provirus in the genome of field strains of fowlpox virus. Complete sequence of the genome of a vaccine-like strain of fowlpox virus has been determined. The functions of the majority of the genes are not known at this time. It is however, interesting that the virus tends to persist in the poultry environment for extended periods of time where other viruses may not survive. In this regard the presence of photolyase gene and A-type inclusion body gene in the virus genome appear to protect the virus from environmental insults. Antigenic cross-reactivity is observed among avianpox viruses and it appears that many genes are conserved. Limited studies on antigenic, genetic and biologic comparison of fowlpox virus with other avianpox viruses especially those that infect the wild birds are available. Recently, complete sequence of canarypox virus genome has become available.

Diagnostic Techniques

Table: Test methods available for diagnosis of fowlpox and their purpose.

Method	Purpose					
	Population freedom from infection	Individual animal freedom from infection prior to movement	Contribution to eradication policies	Confirmation of clinical cases	Prevalence of infection – surveillance	Immune status in individual animals or populations postvaccination

Agent identification						
Virus isolation	–	+	–	+	–	–
Real-time PCR	–	+	–	+	–	–
Detection of immune response						
AGID	–	–	–	–	–	++
ELISA	–	+	–	–	+	++
IFAT	–	–	–	+	–	–

Key: +++ = recommended method; ++ = suitable method; + = may be used in some situations, but cost, reliability, or other factors severely limits its application; – = not appropriate for this purpose.

Although not all of the tests listed as category +++ or ++ have undergone formal validation, their routine nature and the fact that they have been used widely without dubious results, makes them acceptable.

PCR = polymerase chain reaction.

AGID = agar gel immunodiffusion.

ELISA = enzyme-linked immunosorbent assay.

IFAT = indirect fluorescent antibody test.

Identification of the Agent

Fowlpox virus multiplies in the cytoplasm of epithelial cells with the formation of large intracytoplasmic inclusion bodies (Bollinger bodies) that contain smaller elementary bodies (Borrel bodies). The inclusions can be demonstrated in sections of cutaneous and diphtheritic lesions by the use of haematoxylin and eosin (H&E), acridine orange or Giemsa stains. The elementary bodies can be detected in smears from lesions, for example by the Gimenez method, which is described below. Electron microscopy can be used to demonstrate viral particles of typical poxvirus morphology by negative staining or in ultrathin sections of infected tissues.

A Smear Technique for Fowlpox

Test Procedure

- Place a drop of distilled water and the lesion (cutaneous or diphtheritic) on a clean slide. Prepare a thin smear by pressing the lesion with another clean slide and rotating the upper slide several times. Air dry and gently fix the smear over a flame.

- Stain the smear for 5–10 minutes with freshly prepared primary stain (8 ml stock solution2 of basic fuchsin mixed with 10 ml of phosphate buffer3 , pH 7.5, and filtered through Whatman filter paper No. 1).
- Wash thoroughly with tap water.
- Counterstain with malachite green (0.8% in distilled water) for 30–60 seconds. vi) Wash the smear with tap water and then dry. vii) Examine the smear under oil immersion. The elementary bodies appear red and are approximately 0.2–0.3 µm in size.

Virus Isolation

Fowlpox virus can be isolated by the inoculation of suspected material into embryonated chicken eggs or in cell cultures of avian origin. Approximately 0.1 ml of tissue suspension of skin or diphtheritic lesions, treated with the appropriate concentration of antibiotics, is inoculated on to the chorioallantoic membranes (CAMs) of 9- to 12-day-old developing chicken embryos or in cell culture. It is advisable to check the inoculum for any residual contamination by inoculation of a blood agar and McConkey plate examined 24 hours after incubation. Following inoculation of the embryos with the contamination free sample the eggs are incubated at 37 °C for 5–7 days, and then examined for focal white pock lesions or generalised thickening of the CAMs. Histopathological examination of the CAM lesions will reveal eosinophilic intracytoplasmic inclusion bodies following staining with H&E.

Primary chicken embryo fibroblasts, chicken embryo kidney cells, chicken embryo dermis cells, or the permanent quail cell line QT-35, can also be used to propagate fowlpox virus. The adaptation of virus strains to cell cultures is an important requirement for plaque formation, as not all strains will form plaques initially.

Molecular Methods

Restriction fragment length polymorphism (RFLP) analysis can be used for comparison of field isolates and vaccine strains of fowlpox virus. However, this procedure is not used in routine diagnosis.

Cloned genomic fragments of fowlpox virus can be used effectively as nucleic acid probes for diagnosis of fowlpox. Viral DNA isolated from lesions can be detected by hybridisation either with radioactively or nonradioactively labelled genomic probes. This method is especially useful for differentiation of fowlpox from infectious laryngotracheitis when tracheal lesions are present.

Genomic DNA sequences of various sizes can be amplified by the polymerase chain reaction (PCR) using specific primers. This technique is useful when there is only an extremely small amount of viral DNA in the sample. Genomic DNA isolated from the formalinfixed tissue sections from cases that are positive for fowlpox (based upon presence of cytoplasmic inclusions) can be used for PCR amplification of the specific size genomic fragments.

Serological Tests

Although both cell-mediated immunity (CMI) and humoral immunity play an important role in poxvirus infections, routine use of the CMI test is not convenient. Therefore, serological tests, such as virus neutralisation (VN), agar gel immunodiffusion (AGID), passive haemagglutination and fluorescent antibody tests as well as the enzymelinked immunosorbent assay (ELISA), are used to measure specific humoral antibody responses. Evidence of successful immunisation with vaccine can be determined by examining a flock 7–10 days after vaccination for 'takes'. A take consists of a swelling of the skin or a scab at the site where the vaccine was applied and its presence is evidence of successful immunisation.

Virus Neutralisation

After virus/serum interaction, the residual virus activity may be assayed in embryonating chicken eggs or in cell cultures. This technically demanding test may not be convenient for routine diagnosis. Only some selected strains of the virus have plaque-forming ability in chicken embryo cells. Neutralising antibodies develop within 1–2 weeks of infection.

Agar Gel Immunodiffusion

Precipitating antibodies can be detected by reacting test sera against viral antigens. The antigen can be derived by sonication and homogenisation of infected skin or CAM lesions as well as by treatment of infected cell cultures. The lysed suspension is centrifuged and the supernatant is used as antigen. Gel-diffusion medium is prepared with 1% agar, 8% sodium chloride and 0.01% thiomersol. The viral antigen is placed in the central well and the test sera are placed in the peripheral wells. It is important to include a positive and negative control serum. The plates are incubated at room temperature. Precipitation lines develop in 24–48 hours after incubation of the antigen with antibody to homologous or closely related strains. The test is less sensitive than the ELISA or the passive haemagglutination test.

Passive Haemagglutination

Tanned sheep or horse red blood cells are sensitised with a partially purified fowlpox viral antigen. The antigen is prepared from infected CAMs or cells. Passive haemagglutination is more sensitive than AGID. The test will give cross-reactions among avian pox viruses.

Fluorescent Antibody Tests

Direct or indirect immunofluorescence tests will reveal specific intracytoplasmic fluorescence in infected cells. The latter test is commonly used and involves two steps: the antibody against fowlpox virus is reacted with the antigen in the infected cells, followed by a secondary fluoresceinisothiocyanate-labelled antibody against chicken gamma globulin (e.g., goat anti-chicken). Such labelled antibodies are available commercially. In this regard, formalin-fixed tissue sections can be used effectively for fluorescent antibody tests.

Immunoperoxidase

Specific staining of cytoplasmic inclusions is achieved when horseradish-peroxidase-conjugated specific polyclonal antibody against fowlpoxvirus is reacted with the hydrated sections of fowlpoxinfected fixed tissues (CAM and skin) or cell culture. Similar results are obtained when either polyclonal or monoclonal antibodies are used in an indirect test. An advantage of the technique is that the sections can be examined with the light microscope and can be stored for an extended period without loss of colour.

Enzyme-linked Immunosorbent Assay

ELISAs have been developed to detect humoral antibodies to fowlpox virus. They are capable of detecting antibody 7–10 days after infection, but commercial kits for this test are not available.

Fowlpox virus antigens are prepared either from infected QT-35 cell monolayers or CAM lesions. Infected QT-35 cells are pelleted (700 g for 10 minutes at 4 °C), washed with isotonic buffer (10 mM Tris, pH 8.0, 150 mM NaCl, 5 mM ethylene diamine tetra-acetic acid [EDTA]) followed by lysis in hypotonic buffer (10 mM Tris, pH 8.0, 10 mM KCl, 5 mM EDTA) containing 0.1% Triton X-100 and 0.025% beta-mercaptoethanol. Nuclei and cellular debris are removed by low-speed centrifugation (500 g for 5 minutes at 4 °C) and the resulting supernatant is used as a source of fowlpox virus antigens for ELISA or immunoblotting. To isolate viral antigen from CAM lesions, initial grinding of the lesions with subsequent detergent treatment as described earlier would be required. Virus propagated in chicken embryo fibroblasts and chicken embryo dermis cells has also been used for antigen. The antigen preparation is as described for QT-35 cells.

Wells of microtitre plates are coated with 1 μg of soluble fowlpox virus antigen in 100 μl of coating buffer (15 mM Na2CO3, 35 mM NaHCO3, pH 9.6) and incubated overnight at 4 °C. Each well is then rinsed once with wash solution (0.29 M NaCl, 0.05% Tween 20) and then blocked with phosphate buffered saline (PBS, pH 7.4) containing 3% bovine serum albumin (BSA) for 1 hour at 37 °C. After one wash, serial dilutions of the test sera in PBS containing 1% BSA are added to the wells. After rocking for 2 hours at 37 °C, the wells are washed three times prior to the addition of 100 μl/well horseradish-peroxidase-conjugated goat anti-chicken IgY (H + L) antibodies at a recommended dilution in PBS. After 2 hours' incubation at 37 °C and three subsequent washes, 100 μl of the peroxidase substrate TMB3 is added to each well. Reactions are terminated by the addition of 1 M phosphoric acid and absorbance at 450 nm is recorded using an ELISA plate reader.

Immunoblotting

Antigenic variations between strains of fowlpox virus can be evaluated by means of immunoblotting or Western Blotting. In this method, viral antigens separated by SDS-PAGE (sodium dodecyl sulphatepolyacrylamide gel electrophoresis) are reacted either with polyclonal or monoclonal antibodies against fowlpox virus. This method is not convenient for routine diagnosis.

Pseudocowpox Virus

Virion

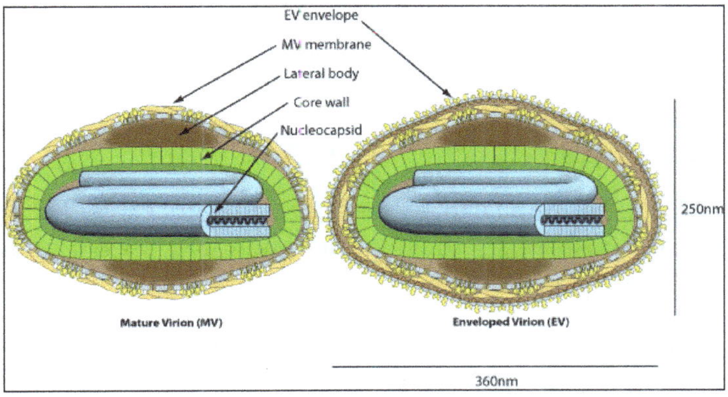

Enveloped, ovoid virion, 220-300nm long and 140-170nm wide. The surface membrane displays surface filaments. Two distinct infectious virus particles exists: the intracellular mature virus (IMV) and the extracellular enveloped virus (EEV).

Genome

Linear, dsDNA genome of 130-150kb. The linear genome is flanked by inverted terminal repeat (ITR) sequences which are covalently-closed at their extremities.

Gene Expression

Replication: Cytoplasmic

- Attachment of the viral proteins to host glycosaminoglycans (GAGs) mediates endocytosis of the virus into the host cell.
- Fusion with the plasma membrane to release the core into the host cytoplasm.
- Early phase: Early genes are transcribed in the cytoplasm by viral RNA polymerase. Early expression begins at 30 minutes post-infection.

- Core is completely uncoated as early expression ends, viral genome is now free in the cytoplasm.
- Intermediate phase: Intermediate genes are expressed, triggering genomic DNA replication at approximately 100 minutes post-infection.
- Late phase: Late genes are expressed from 140 min to 48 hours post-infection, producing all structural proteins.
- Assembly of progeny virions starts in cytoplasmic viral factories, producing an spherical immature particle. This virus particle matures into brick-shaped intracellular mature virion (IMV).
- IMV virion can be released upon cell lysis, or can acquire a second double membrane from *trans*-Golgi and bud as external enveloped virion (EEV).

Asafarviradae

This disease can affect pigs of any age. It has high mortality rates. The key clinical signs include blue-purple cyanosis of snout, ears, tail and lower legs; high fever; and heavy discharge from eyes and nose.

African swine fever (ASF) resembles classical swine fever (CSF) (hog cholera) so closely that laboratory tests are required to differentiate them. The clinical signs and post-mortem lesions of the two diseases are almost indistinguishable. ASF is caused by a unique virus which is distinct from that of CSF and which infects only domestic and wild pigs and a variety of soft bodied ticks. The virus is endemic in Africa south of the equator, in warthogs and bush pigs, but the infection in them produces no clinical disease. It circulates between warthogs and the soft bodied ticks which inhabit their burrows. The ticks transmit it through all stages of their life cycle and perpetuate it. It is also endemic in the domestic pigs of some African countries.

The pig (and its close relatives, boars and hogs) is the only natural host of the double-stranded, *Asfarviridae* family of viruses, meaning the virus does not cause harm to humans or other animals. This does not mean that humans and other animals cannot spread the virus as carriers; African swine fever (ASF) is commonly carried by arthropods, such as the soft-bodied tick, through uptake of blood from infected pigs.

Contamination generally occurs via direct contact with tissue and bodily fluids from infected or carrier pigs, including discharges from the nose, mouth, urine and faeces or infected semen. It also spreads through transport and consumption of contaminated food products, and some cases have originated from failure to comply with biosecurity standards by feeding waste food to domestic pigs. It is believed that a highly pathogenic strain of ASF was introduced to domestic pigs and, subsequently, wild boar populations in the port of Poti, Georgia, in 2007 when waste food from a ship originating in South Africa was fed to local pigs.

Although the virus in wild boar and hogs does not manifest any signs of the disease, it remains highly contagious across all swine species and can survive in pigs for long periods of time post-slaughter – even in frozen carcasses. It is also important to note that curing and smoking pork products does not destroy the virus.

It is vital to immediately distinguish the disease that is infecting a herd; ASF and classical swine fever are caused by very similar viruses which are only distinguishable by laboratory testing. Notifying a vet as soon as any signs arise is the best way to ensure the correct quarantine and treatment procedures are followed – it could save the rest of your pigs.

Clinical Signs

- High fever 40-42 °C.
- Loss of appetite.
- Depression.
- Lethargic – sometimes refusal to stand or move.
- Very unsteady when stood up.
- Vomiting and diarrhoea with bloody discharge.
- White skinned pigs: Extremities (nose, ears, tail and lower legs) become cyanotic (blue-purple colour).
- Discrete haemorrhages appear in the skin particularly on the ears and flanks.
- Group will huddle together and are usually shivering.
- Abnormal breathing.
- Heavy discharge from eyes and nose.
- Comatose state and death within a few days.
- Some pigs can show conjunctivitis with reddening of the conjunctival mucosa and ocular discharges.

Pregnant sows commonly undergo miscarriage or deliver stillborn piglets that are malformed – piglets can be tested for the virus.

Mortality rate in infected groups of pigs is high and there is no vaccination proven to prevent or cure infection, therefore, it is crucial that control begins on-farm. European, South American and Caribbean countries which have been infected have adopted a slaughter policy to eradicate the virus within the herd. Mild strains of the virus also occur which cause a milder but equally serious disease in domestic pig herds – individuals from these herds must also be slaughtered to prevent pathogenesis.

Diagnosis

Pigs that die early in an outbreak may not have any noticeable lesions, but as the disease progresses the lesions become striking. Bright red haemorrhages in the lymph nodes, kidneys, heart and linings of the body cavities are common findings. There may also be excess haemorrhagic fluid in the body cavities and gelatinous fluid in the lungs. The spleen may be enlarged, darkened and crumble on slight pressure.

The veterinarian will have to send samples to a laboratory which specialises in CSF and ASF diagnosis. The best samples to send are blood, lymph nodes, spleen and, in chronic cases, serum for serology. In the case it is CSF and not ASF, the tonsils might also be sent. The veterinarian should consult the appropriate veterinary authorities on how best to send these.

The tonsils of the pig are very easy to find. Laying the dead pig on its back, cut away the skin and flesh under and between its lower jaw bone and tongue. The pair of tonsils are two large red patches each about the size of the end half of your thumb or perhaps slightly bigger. Their surfaces are covered with small pits or depressions.

In South Africa and countries outside Africa it is essential to isolate and identify the virus. Only about six laboratories in the world can do this. In African countries where the disease is endemic in the domestic pig population, the veterinarian may only send serum samples for antibody detection.

The virus may be isolated in primary cultures of pig bone marrow or peripheral blood leucocytes. Infected cells haemadsorb i.e., pig red cells will adhere to them. Virus can also be detected in infected cells by fluorescent antibody tests. ELISA tests are also used to detect antibodies. In doubtful cases samples can be injected into experimental pigs.

Serum antibody titres may be tested in a number of ways. The indirect immunofluorescence (IIF) and the ELISA tests seem to be the most favoured.

That porcine dermatitis and nephropathy syndrome, which occurs from time to time in most pig rearing areas can resemble ASF and CSF clinically and at post-mortem examination. Laboratory examination may be necessary to eliminate them form the diagnosis.

Cause

African swine fever is caused by the *Asfarviridae* family of viruses which are distinct from the viruses associated with Classical swine fever. There are 22 known types of the ASF virus, allowing the epidemiological tracing of outbreaks to the source.

The infection can be introduced to uninfected herds in a number of ways:

- The feeding of contaminated feed and contaminated food waste used to supplement feed.
- Through the bites of soft-bodied ticks, lice and flies.
- Through inoculation with contaminated syringes and use of contaminated surgical equipment.
- Through the introduction of infected pigs to the herd.

Transmission of the virus within the herd is generally through direct contact with infected bodily discharges, faeces and vomit.

Prevention

There is no live or attenuated vaccine for the prevention of ASF therefore control of the virus is reliant on strict biosecurity.

- Do not feed domestic pigs food waste; this is illegal in the UK, other EU regions and some states within the US.
- Where 'permitted garbage feeding' is legal in US states, pigs fed this way are prohibited from exportation.
- Do not leave food waste exposed for wild swine species to access. Dispose of food waste properly.
- Abide by strict biosecurity rules. Do not take pig meat onto farms, or restrict all food (and consumption of food) to a canteen. All staff on farm should be inducted onto a strict programme of hand and equipment sanitisation before and after contact with pigs.
- Follow rules and regulations on disposal of food waste at ferry ports and airports.
- Provide the means for staff and visitors to thoroughly sanitise their hands and equipment.
- Ensure that wild boar, warthogs and wild pigs, and materials potentially contaminated by such wild species do not come into contact with domestic pigs.
- Check infected regions before import of goods that could potentially be contaminated.
- Advise and educate people on the risks of bringing back pork products from infected regions.

Iridoviridae

Virions display icosahedral symmetry and contain an internal lipid membrane located between the DNA–protein core and the outer capsid. Depending upon the genus, the capsid may have numerous external fibrils. Mature infectious virions may remain non-enveloped or bud from the plasma membrane and acquire an envelope.

Table: Characteristics of the family Iridoviridae.

Typical member:	Frog virus 3 (AY548484), species *Frog virus 3*, genus *Ranavirus*.
Virion	Typically 150–200 nm (non-enveloped); the principal component of the capsid is the major capsid protein (mol wt 48 kDa).
Genome	Linear, double-stranded circularly permuted, terminally redundant DNA, 103–220 kbp, encoding 92–211 proteins.
Replication	First-stage DNA synthesis and early transcription takes place in the nucleus; subsequently DNA concatemer formation and late transcription occur in the cytoplasm; virion morphogenesis takes place in cytoplasmic assembly sites.
Translation	Directly from capped, non-polyadenylated mRNAs.
Host Range	Amphibians, reptiles, fish (subfamily *Alphairidovirinae*); mainly insects and crustaceans (subfamily *Betairidovirinae*).
Taxonomy	Five genera divided between two subfamilies.

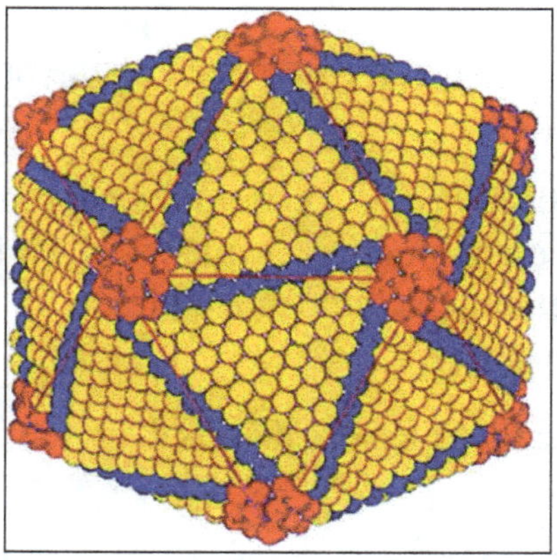

Proposed structure of the capsid of invertebrate iridescent virus 2.
Trisymmetrons (orange) comprising the icosahedral faces, pentasymmetrons (red) located at the vertices, and disymmetrons (blue) at the edges of the faces are shown.

Genome

The virus genome is a single molecule of double-stranded DNA that is circularly permuted and terminally redundant. The degree of redundancy varies from 5–50% depending on the virus species. The genomes of all vertebrate iridoviruses, with one exception, are highly methylated due to a virus-encoded DNA cytosine methyltransferase. Twenty-six core proteins are common to all members of the family.

Replication

Replication involves both nuclear and cytoplasmic compartments. In the nucleus, host RNA polymerase II directs the synthesis of immediate–early virus mRNAs. In addition a virus-encoded DNA polymerase synthesizes genome- to twice-genome-size copies of the incoming DNA. This DNA is transported to the cytoplasm, where it undergoes a second round of replication to form large concatemers. Late gene expression is catalysed by a virus-encoded transcriptase. Virion morphogenesis occurs within assembly sites, and completed particles form paracrystalline arrays or bud from the plasma membrane.

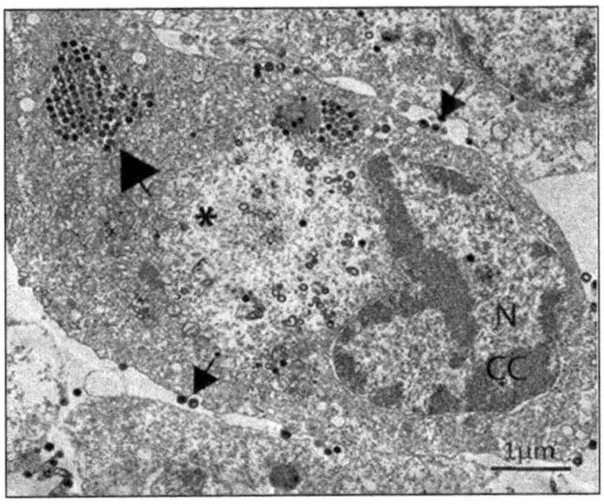

Transmission electron micrograph of a fathead minnow cell infected with frog virus 3. Large arrow, paracrystalline array; small arrows, budding virions; N, nucleus displaying condensed chromatin (CC) indicative of apoptosis; *, virus assembly site showing empty and complete virus particles; scale bar = 1 µm.

Taxonomy

Phylogenetic analyses indicate that invertebrate and vertebrate iridovirus lineages diverged early during the evolution of the family. Ascoviruses appear to have emerged recently from the former lineage and share with iridoviruses a common ancestry with marseilleviruses.

Ranavirus

Ranaviruses are promiscuous pathogens capable of infecting three classes of ectothermic vertebrates (bony fish, amphibians and reptiles). Ranaviruses infect not only multiple species within a class, but some (e.g. Bohle iridovirus) are also capable of infecting hosts from different classes. Infections are systemic, involve multiple internal organs and may lead to high levels of morbidity and mortality among cultured, commercially important fish and amphibians, as well as endangered wild species.

Megalocytivirus

Members infect >50 species of marine and freshwater fish; systemic disease involves multiple internal organs.

Lymphocystivirus

Lymphocystiviruses infect >100 species of marine and freshwater fish, leading to the formation of wart-like growths composed of clusters of individual, infected cells (some as large as 1 mm) primarily on the skin, but sometimes on internal organs. Morbidity may be high but mortality tends to be low.

Iridovirus and Chloriridovirus

Members of these genera infect >100 insect and crustacean species. Patent infections involve massive levels of virus replication that result in infected larvae displaying marked iridescence, whereas covert infections may reduce the reproductive capacity of the host. The genera were previously distinguished based on virion size and the iridescent colour of infected larvae; phylogenetic analysis of complete genome sequences now gives a superior method of differentiation.

References

- Ectromelia-virus: microbewiki.kenyon.edu, Retrieved 01 June, 2019
- Rabbitpox, veterinary-science-and-veterinary-medicine: sciencedirect.com, Retrieved 19 July, 2019
- Pox, swine-index-diseases: vetmed.iastate.edu, Retrieved 17 May, 2019
- Shope-fibroma-virus, rabbit-condition: wagwalking.com, Retrieved 16 March, 2019
- African-swine-fever-asf, disease-guide: thepigsite.com, Retrieved 14 April, 2019
- Infectious Diseases of Livestock, 2nd Edition. Oxford University Press

3

Animal Herpesvirales Virus

Herpesvirales is a dsDNA virus which is recognized by icosahedral capsid enclosed in a lipid envelope. Some of its types include bovine herpesvirus, equine herpesvirus type 1, felid herpesvirus, suid herpesvirus, duck virus enteritis, oncorhynchus masou virus, etc. These diverse viruses and diseases in animals related to herpesvirales virus have been thoroughly discussed in this chapter.

Herpesvirales Virus

Herpesviridae is a large family of DNA viruses that cause infections and certain diseases in animals, including humans. The members of this family are also known as herpesviruses. The family name is derived from the word *herpein* ("to creep"), referring to spreading cutaneous lesions, usually involving blisters, seen in flares of herpes simplex 1, herpes simplex 2 and herpes zoster (shingles). In 1971, the International Committee on the Taxonomy of Viruses (ICTV) established Herpesvirus as a genus with 23 viruses among four groups. Latent, recurring infections are typical of this group of viruses, though the family name does not refer to latency. *Herpesviridae* can cause latent or lytic infections.

At least five species of the *Herpesviridae* – HSV-1 and HSV-2 (both of which can cause orolabial herpes and genital herpes), varicella zoster virus (the cause of chickenpox and shingles), Epstein–Barr virus (implicated in several diseases, including mononucleosis and some cancers), and cytomegalovirus – are extremely widespread among humans. More than 90% of adults have been infected with at least one of these, and a latent form of the virus remains in almost all humans who have been infected.

Nine herpesvirus types are known to infect humans: herpes simplex viruses 1 and 2 (HSV-1 and HSV-2, also known as HHV1 and HHV2), varicella-zoster virus (VZV, which may also be called by its ICTV name, HHV-3), Epstein–Barr virus (EBV or HHV-4), human cytomegalovirus (HCMV or HHV-5), human herpesvirus 6A and 6B (HHV-6A and HHV-6B), human herpesvirus 7 (HHV-7), and Kaposi's sarcoma-associated herpesvirus (KSHV, also known as HHV-8). In total, more than 130 herpesviruses are known, some of them from mammals, birds, fish, reptiles, amphibians, and mollusks.

Structure

All members of the *Herpesviridae* share a common structure; a relatively large, monopartite, double-stranded, linear DNA genome encoding 100-200 genes encased within an icosahedral protein cage (with T=16 symmetry) called the capsid, which is itself wrapped in a protein layer called the tegument containing both viral proteins and viral mRNAs and a lipid bilayer membrane called the envelope. This whole particle is known as a virion.

Life Cycle

All herpesviruses are nuclear-replicating—the viral DNA is transcribed to mRNA within the infected cell's nucleus.

Infection is initiated when a viral particle contacts a cell with specific types of receptor molecules on the cell surface. Following binding of viral envelope glycoproteins to cell membrane receptors, the virion is internalized and dismantled, allowing viral DNA to migrate to the cell nucleus. Within the nucleus, replication of viral DNA and transcription of viral genes occurs.

During symptomatic infection, infected cells transcribe lytic viral genes. In some host cells, a small number of viral genes termed latency-associated transcript (LAT) accumulate, instead. In this fashion, the virus can persist in the cell (and thus the host) indefinitely. While primary infection is often accompanied by a self-limited period of clinical illness, long-term latency is symptom-free.

Chromatin dynamics regulate the transcription competency of entire herpes virus genomes. When the virus enters a cell, the cellular immune response is to protect the cell. The cell does so by wrapping the viral DNA around histones and condensing it into chromatin, causing the virus to do dormant, or latent. If cells are unsuccessful and the chromatin is loosely bundled, the viral DNA is still accessible. The viral particles can turn on their genes and replicate using cellular machinery to reactivate, starting a lytic infection.

Reactivation of latent viruses has been implicated in a number of diseases (e.g. shingles, pityriasis rosea). Following activation, transcription of viral genes transitions from LAT to multiple lytic genes; these lead to enhanced replication and virus production. Often, lytic activation leads to cell death. Clinically, lytic activation is often accompanied by emergence of nonspecific symptoms, such as low-grade fever, headache, sore throat, malaise, and rash, as well as clinical signs such as swollen or tender lymph nodes and immunological findings such as reduced levels of natural killer cells.

In animal models, local trauma and system stress has been found to induce reactivation of latent herpesvirus infection. Cellular stressors like transient interruption of protein synthesis and hypoxia are also sufficient to induce viral reactivation.

Genus	Host details	Tissue tropism	Entry details	Release details	Replication site	Assembly site	Transmission
Iltovirus	Birds: galliform: psittacine	None	Cell receptor endocytosis	Budding	Nucleus	Nucleus	Oral-fecal, aerosol
Proboscivirus	Elephants	None	Glycoproteins	Budding	Nucleus	Nucleus	Contact
Cytomegalovirus	Humans; monkeys	Epithelial mucosa	Glycoproteins	Budding	Nucleus	Nucleus	Urine, saliva
Mardivirus	Chickens; turkeys; quail	None	Cell receptor endocytosis	Budding	Nucleus	Nucleus	Aerosol
Rhadinovirus	Humans; mammals	B-lymphocytes	Glycoproteins	Budding	Nucleus	Nucleus	Sex, saliva
Macavirus	Mammals	B-lymphocytes	Glycoproteins	Budding	Nucleus	Nucleus	Sex, saliva
Roseolovirus	Humans	T-cells; B-cells; NK-cell; monocytes; macrophages; epithelial	Glycoproteins	Budding	Nucleus	Nucleus	Respiratory contact
Simplexvirus	Humans; mammals	Epithelial mucosa	Cell receptor endocytosis	Budding	Nucleus	Nucleus	Saliva
Scutavirus	Sea turtles	None	Cell receptor endocytosis	Budding	Nucleus	Nucleus	Aerosol
Varicellovirus	Mammals	Epithelial mucosa	Glycoproteins	Budding	Nucleus	Nucleus	Aerosol
Percavirus	Mammals	B-lymphocytes	Glycoproteins	Budding	Nucleus	Nucleus	Sex, saliva
Lymphocryptovirus	Humans; mammals	B-lymphocytes	Glycoproteins	Budding	Nucleus	Nucleus	Saliva
Muromegalovirus	Rodents	Salivary glands	Glycoproteins	Budding	Nucleus	Nucleus	Contac

Evolution

The three mammalian subfamilies – Alpha-, Beta- and Gamma-herpesviridae – arose approximately 180 to 220 mya. The major sublineages within these subfamilies were probably generated before the mammalian radiation of 80 to 60 mya. Speciations within sublineages took place in the last 80 million years probably with a major component of cospeciation with host lineages.

All the currently known bird and reptile species are alphaherpesviruses. Although the branching order of the herpes viruses has not yet been resolved, because herpes viruses and their hosts tend to coevolve this is suggestive that the alphaherpesviruses may have been the earliest branch.

The time of origin of the genus Iltovirus has been estimated to be 200 mya while those of the mardivirus and simplex genera have been estimated to be between 150 and 100 mya.

Immune System Evasions

Herpesviruses are known for their ability to establish lifelong infections. One way this is possible is through immune evasion. Herpesviruses have many different ways of evading the immune system. One such way is by encoding a protein mimicking human interleukin 10 (hIL-10) and another is by downregulation of the major histocompatibility complex II (MHC II) in infected cells.

CmvIL-10

Research conducted on cytomegalovirus (CMV) indicates that the viral human IL-10 homolog, cmvIL-10, is important in inhibiting pro-inflammatory cytokine synthesis. The cmvIL-10 protein has 27% identity with hIL-10 and only one conserved residue out of the nine amino acids that make up the functional site for cytokine synthesis inhibition on hIL-10. There is, however, much similarity in the functions of hIL-10 and cmvIL-10. Both have been shown to down regulate IFN-γ, IL-1α, GM-CSF, IL-6 and TNF-α, which are all pro-inflammatory cytokines. They have also been shown to play a role in downregulating MHC I and MHC II and up regulating HLA-G (non-classical MHC I). These two events allow for immune evasion by suppressing the cell-mediated immune response and natural killer cell response, respectively. The similarities between hIL-10 and cmvIL-10 may be explained by the fact that hIL-10 and cmvIL-10 both use the same cell surface receptor, the hIL-10 receptor. One difference in the function of hIL-10 and cmvIL-10 is that hIL-10 causes human peripheral blood mononuclear cells (PBMC) to both increase and decrease in proliferation whereas cmvIL-10 only causes a decrease in proliferation of PBMCs. This indicates that cmvIL-10 may lack the stimulatory effects that hIL-10 has on these cells.

It was found that cmvIL-10 functions through phosphorylation of the Stat3 protein. It was originally thought that this phosphorylation was a result of the JAK-STAT pathway. However, despite evidence that JAK does indeed phosphorylate Stat3, its inhibition has no significant influence on cytokine synthesis inhibition. Another protein, PI3K, was also found to phosphorylate Stat3. PI3K inhibition, unlike JAK inhibition, did have a significant impact on cytokine synthesis. The difference between PI3K and JAK in Stat3 phosphorylation is that PI3K phosphorylates Stat3 on the S727 residue whereas JAK phosphorylates Stat3 on the Y705 residue. This difference in phosphorylation positions seems to be the key factor in Stat3 activation leading to inhibition of pro-inflammatory cytokine synthesis. In fact, when a PI3K inhibitor is added to cells, the cytokine synthesis levels are significantly restored. The fact that cytokine levels are not completely restored indicates there is another pathway activated by cmvIL-10 that is inhibiting cytokine system synthesis. The proposed mechanism is that cmvIL-10 activates PI3K which in turn activates PKB (Akt). PKB may then activate mTOR, which may target Stat3 for phosphorylation on the S727 residue.

MHC Downregulation

Another one of the many ways in which herpes viruses evade the immune system is by down regulation of MHC I and MHC II. This is observed in almost every human herpesvirus. Down regulation of MHC I and MHC II can come about by many different mechanisms, most causing the MHC to be absent from the cell surface. As discussed above, one way is by a viral chemokine homolog such as IL-10. Another mechanism to down regulate MHCs is to encode viral proteins that detain the newly formed MHC in the endoplasmic reticulum (ER). The MHC cannot reach the cell surface and therefore cannot activate the T cell response. The MHCs can also be targeted for destruction in the proteasome or lysosome. The ER protein TAP also plays a role in MHC down regulation. Viral proteins inhibit TAP preventing the MHC from picking up a viral antigen peptide. This prevents proper folding of the MHC and therefore the MHC does not reach the cell surface.

It is important to note that HLA-G is often up regulated in addition to downregulation of MHC I and MHC II. This prevents the natural killer cell response.

Zoonotic Herpesviruses

In addition to the herpesviruses considered endemic in humans, some viruses associated primarily with animals may infect humans. These are zoonotic infections:

Species	Type	Synonym	Subfamily	Human Pathophysiology
Macaque monkey	CeHV-1	Cercopithecine herpesvirus 1, (monkey B virus)	α	Very unusual, with only approximately 25 human cases reported. Untreated infection is often deadly; sixteen of the 25 cases resulted in fatal encephalomyelitis. At least four cases resulted in survival with severe neurologic impairment. Symptom awareness and early treatment are important for laboratory workers facing exposure.
Mouse	MuHV-4	Murid herpesvirus 68 (MHV-68)	γ	Zoonotic infection found in 4.5% of general population and more common in laboratory workers handling infected mice. ELISA tests show factor-of-four (x4) false positive results, due to antibody cross-reaction with other Herpes viruses.

Animal Herpesviruses

In animal virology, the best known herpesviruses belong to the subfamily *Alphaherpesvirinae*. Research on pseudorabies virus (PrV), the causative agent of Aujeszky's disease in pigs, has pioneered animal disease control with genetically modified vaccines. PrV is now extensively studied as a model for basic processes during lytic herpesvirus infection, and for unraveling molecular mechanisms of herpesvirus neurotropism, whereas bovine herpesvirus 1, the causative agent of bovine infectious rhinotracheitis and pustular vulvovaginitis, is analyzed to elucidate molecular mechanisms of latency. The avian infectious laryngotracheitis virus is phylogenetically distant from these two viruses and serves to underline similarity and diversity within the *Alphaherpesvirinae*.

Bovine Herpesvirus

Bovine herpesvirus 1 (BoHV-1), the causative agent of infectious bovine rhinotracheitis, is a well-studied pathogen that infects cattle in many parts of the world. An infected animal can exhibit a wide variety of symptoms depending on the specific conditions of infection, including respiratory problems, lesions, and increased abortion rates. BoHV-1 can be transmitted through various means, including the exchange of nasal discharge, semen transmission, and aerosol transmission. This makes the virus highly contagious, especially in a crowded feedlot environment. BoHV-1 is especially problematic because of its ability to enter a latency phase within the host, meaning that an apparently healthy animal might suddenly resume the virus's spread.

Although infection by BoHV-1 is not usually lethal, it does detract from the productivity of the animal. If the infection is not contained and is allowed to spread to the entire herd, there will be significant economic losses incurred from deaths, abortions, reduced body mass, reduced dairy production, and the cost of treatment. Infection by BoHV-1 is also one of the many contributing factors that can lead to the development of the general disease complex Bovine Respiratory Disease (BRD), which is estimated to cost the United States cattle industry about 640 million dollars annually.

Genome Structure

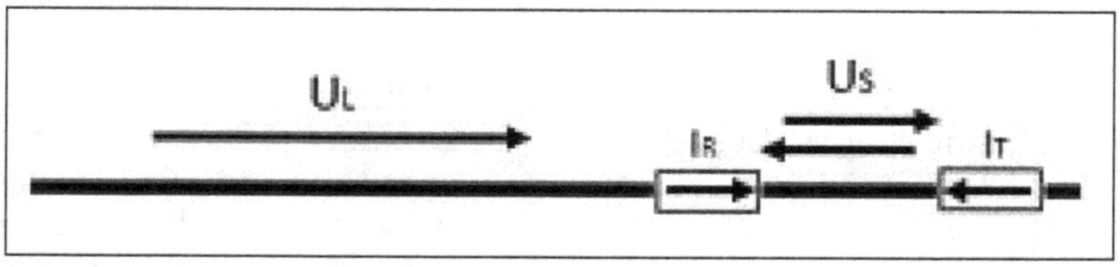

A diagram illustrating the basic organization of the class D BoHV-1 genome. UL and US correspond to unique genome regions. The IR and IT regions are inverted copies of the same sequence, which flank the US segment. The US segment can be inverted, which means that there are two common isomeric forms of the genome.

Like all herpesviruses, BoHV-1 has a relatively long, double stranded DNA genome. Although packaged linearly, it is thought to circularize during its infectious cycle prior to genome replication. It is 135,300 base pairs in length, has a 72% guanine/ cytosine content, and has 73 open reading frames (ORFs). Of these 73 ORFs, 33 have been determined to be essential for viral replication. Most of the virus's genes have homologous counterparts common to all alphaherpesviruses, although some are exclusive to the Varicelloviruses. Only one of these genes (UL0.5) is specific to BoHV-1 alone.

BoHV-1 has a class D herpesvirus genome, which means that its genome contains two unique regions, one long and one short (UL and US), the second of which is flanked by inverted repeats (IR and TR). Both of these unique regions can be inverted, meaning that there are actually four isomeric forms of the herpesvirus genome that can be found in a BoHV-1 virion. In reality, one specific orientation of UL is strongly preferred, so only two isomeric forms of the genome are commonly found in the environment (one corresponding to each inversion of US). These inverted sequences are likely attributable to recombination during genome replication.

Virion Structure of Bovine Herpesvires 1

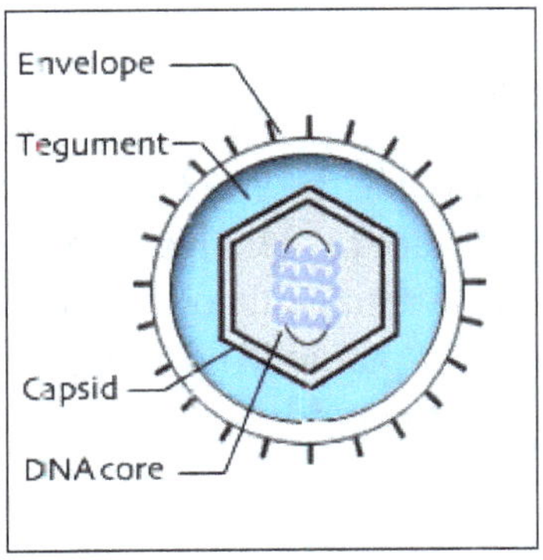

A simple diagram of a typical herpesvirus virion.

BoHV-1 has a typical herpesvirus virion structure: the virus's double stranded DNA genome is contained within an icosahedral protein capsid. The capsid is wrapped in a protein complex called the tegument, which is made up of about 20 viral proteins. The tegument connects the capsid with the outer cell-derived envelope, which contains the viral membrane proteins and glycoproteins that are essential for the successful penetration of the cell membrane, including glycoproteins gD, gB, gH, and gL.

Reproductive Cycle of Bovine Herpesvirus 1 in a Host Cell

The BoHV-1 virion penetrates the cell membrane via a three step process. First, glycoproteins gB and gC on the virion envelope interact with certain cellular structures, creating a low affinity attachment between the virus and the host cell. Second, glycoprotein gD binds to cell membrane protein nectin-1, an immunoglobulin protein. This initiates the third phase, during which the virion envelope fuses with the cell membrane, allowing the capsid and tegument to enter the cytoplasm.

Upon entering the host cell, the virion begins to move towards the nucleus. As it is transported, the tegument proteins surrounding the capsid are shed into the cytoplasm. Although many of these tegument proteins are poorly understood, some are known to have important functions, such as disabling host defenses or subverting the host's resources. For example, the virion host shutoff (vhs) tegument protein is responsible for halting the host's regular protein synthesis, and tegument protein VP16 is required to induce expression of early BoHV-1 genes.

Once it has breached the nuclear membrane, the linear viral genome end-joins and becomes circular, and a combination of viral and cellular proteins induces the expression of the "immediate early" (IE) genes. The IE gene products induce the expression of "early" genes, at which point viral DNA replication begins. "Early late" gene expression begins during viral genome replication, and then finally the "true late" genes are expressed, which code for the capsid-forming structural proteins. Meanwhile, replication of the now-circular viral genome proceeds via a "rolling circle" mechanism, which produces multiple genomes connected to each other in sequence from head to tail (concatemeric DNA). This long strand of DNA is then cleaved into individual copies of the viral genome, which are loaded into the newly-formed virion capsids without leaving the nucleus.

How the virion capsid proceeds to exit the nucleus and acquire its tegument proteins and outer envelope is a matter of debate. The leading theory suggests that the mature capsid acquires a primary envelope as it buds out of the inner nuclear membrane, and into the perinuclear space. The new primary membrane then fuses with the outer nuclear membrane, so that the naked capsid is released into the cytoplasm. The capsid then acquires its tegument coating and final envelope by budding with a trans Golgi compartment.

Viral Ecology and Pathology

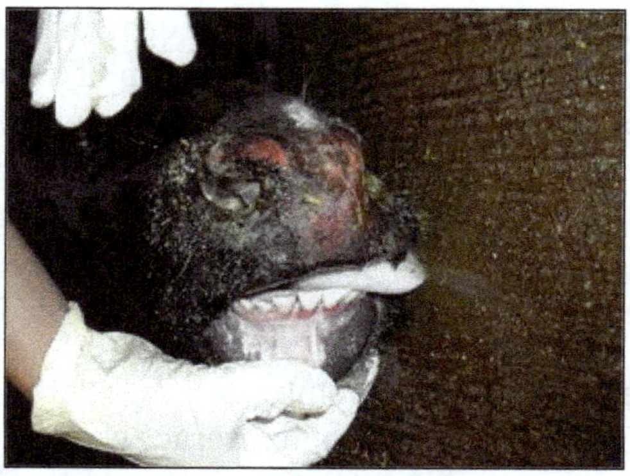

A cow that has been infected by BoHV-1. The animal's excessive nasal discharge is typical of BoHV-1 infection, and is an effective means for the virus to be transmitted to a new host.

Cattle are very commonly kept in extremely crowded and unsanitary conditions, which allows BoHV-1 to easily spread through a herd. The most common route of infection by BoHV-1 is via the direct exchange of nasal runoff between an infected individual and an uninfected individual, although other routes of horizontal transmission include genital contact, semen transmission, and short-distance aerosol transmission. The virus can also be transmitted vertically from mother to calf during pregnancy.

The virus immediately infects the new host's epithelial cells at the contact site, where it rapidly begins its standard lytic replication cycle, spreading to surrounding cells via local dissemination. If the virus makes contact with the ordinary infection site, the nasal epithelial cells, then it can very quickly establish secondary infections via the shedding of virion-loaded nasal mucus.

After spreading within the same tissue, the virus may go on to infect the host's blood, although the mechanism of this advance is poorly understood. Access to the circulatory system gives the virus access to many other types of the host's tissue, accounting for the large diversity in symptoms. In addition to this systemic spread, the virus can also make contact with nerve endings in nasal mucosal surfaces, through which the infection spreads to the central nervous system. BoHV-1 typically infects the trigeminal nerve endings, through which it establishes a latent infection in ganglionic neurons. Even after the animal stops exhibiting symptoms of infection, the latent infection can be reactivated later in life, at which point the animal begins shedding the virus once more.

Because BoHV-1 has significant economic consequences for the beef and dairy industries, considerable efforts have been made to inhibit the virus's spread. Multiple vaccines have been developed for BoHV-1, most of which use an inactivated or attenuated form of the virus. While these vaccines are effective at preventing the onset of symptoms, they do not fully prevent the infection, and the virus can still go on to establish a latent infection. Research into the development of more advanced vaccines is ongoing.

Equine Herpesvirus Type 1

Cause

EHV-1 virus is one of a large group of DNA viruses and is an important cause of disease in horses. EHV-1 causes both sporadic and epidemic abortion (abortion storms), as well as respiratory disease and neurological disease. The neurological strain of EHV-1 is very rare in Australia.

EHV-1 is a common cause of respiratory disease of horses. Once infected, it is assumed horses carry the virus for life. When horses are stressed, the virus can be reactivated and excreted.

Affected Animals

- Horses,
- Donkeys.

Symptoms

Symptoms of both EHV-1 include:

- Fever (39-40.5°C),
- Conjunctivitis,
- Depression,
- Nasal discharge and a cough,
- Lack of appetite,
- Possible swelling of the lymph glands around the throat.

Mild respiratory disease may involve only conjunctivitis and a cough with little impairment of respiratory function. The nasal discharge and cough can last up to 3 weeks. Secondary bacterial infection can result in pneumonia.

Abortion in horses due to EHV-1 usually occurs between 8-10 months' gestation and, occasionally, as early as 4-5 months' gestation. Abortions occur anywhere from 10 days to 12 weeks after virus infection. The mare may not show signs of respiratory infection prior to aborting and sometimes the foetus is expelled still covered with the placenta.

Mares infected late in pregnancy may have a live foal. Infected foals may be normal at birth, but become weak, very depressed and die in a few days with signs of respiratory disease.

Horses with neurological disease caused by EHV-1 infection show signs of incoordination, weakness and have trouble standing. Rear limbs tend to be more severely affected than forelimbs. Other signs can include difficulty urinating and defaecating, extreme lethargy and a coma-like state.

Impacts

EHV-1 infection is likely to result in loss of training time and reduced performance of working or racing horses. Losses due to abortion and peri-natal deaths may be significant, particularly during abortion storms.

How it is Spread

This disease is highly contagious and spreads easily in horse populations by direct horse-to-horse contact or by contaminated fomites such as equipment (e.g. feed and water buckets) and tack (e.g. halters, bridles).

Infection of EHV-1 occurs mainly by inhalation, but also by ingestion of material contaminated by nasal discharges or aborted foetuses.

Monitoring and Action

Actions if an Aborted Foal is Found or Abortions are Suspected

- Isolate the mare from contact with all other horses.
- Leave her halter and lead rope with her in isolation.
- Call your veterinarian. Your veterinarian will provide you with advice on determining the cause and biosecurity measures to take to protect other horses. Your veterinarian will also provide advice on notifying Biosecurity Queensland.
- Use a disposable respirator and wear gloves to pick up the foetus and placenta and put them in a strong plastic bag.
- Keep the foetus and placenta cool (not frozen) for veterinary examination:
    - When aborted foetuses are necropsied, gross foetal lesions may include subcutaneous oedema, jaundice, increased thoracic fluid volume and an enlarged liver with yellow-white lesions.
    - Foetal membranes and a range of foetal tissues, both preserved and either fresh chilled or in virus transport medium should be submitted to the laboratory for diagnostic testing. Lung, liver, adrenal glands and lymphatic tissues should be provided.
    - In cases where EHV-1 abortion is suspected but the foetus and foetal membranes cannot be found, paired serum samples from the mare may help determine if EHV-1 was the cause of abortion.
- Keep other horses away from the abortion site for at least two weeks after the area has been decontaminated.

Anyone who has had contact with the aborted mare or the foetus or placenta should:

- Disinfect hands and boots with a good surface disinfectant, for example iodophors chlorhexidine or phenols.
- Shower and change personal clothing. Washing and sundrying clothes will effectively inactivate the virus.
- Disinfect vehicles and floats used to transport the aborted mare and anything that has come in contact with the foetus, foetal fluids or discharge from the mare.

Until a diagnosis is reached:

- Keep the mare completely isolated.
- Avoid entering the contaminated area at all if possible or use a disinfectant foot-bath outside the area to disinfect footwear.
- Provide the isolated mare with her own feed bin and other equipment, and do not use them for any other horse.
- Make sure the isolated mare is cared for by someone who does not work with other horses or who disinfects, showers and changes clothes before they do.

Prevention

Disease introduction can be prevented and disease spread restricted by good biosecurity and management practices:

- Keep weanlings, yearlings and non-breeding stock in a separate part of the farm with their own facilities and staff.
- Separate pregnant mares from non-pregnant mares and from mares with foals.
- Keep resident mares separated from mares visiting only for the stud season.
- Isolate visiting mares on arrival at stud for 3-4 weeks. Keep them in small groups of 2 or 3, which can be combined if they remain disease-free.
- Minimise stress factors for pregnant mares such as overcrowding and transport in the last 2 months of gestation, and provide access to good quality feed and ensure adequate parasite control.
- Establish isolation facilities equipped with gloves, plastic bags, disinfectant etc.
- Clean and disinfect horse floats after each journey.
- Train staff so they understand about isolation and the principles of preventing disease spread. Implement hygiene procedures that will help prevent spread of disease.
- Keep accurate records of mares (e.g. date of arrival, where from, previous stud details) and of movements of horses on the farm.
- Talk to your veterinarian about vaccinating pregnant mares to help reduce the risk of abortion.

Treatment

There is no specific treatment for EHV1 infection in horses.

Control

- Isolate mares that have aborted for at least 30 days.
- Consult your veterinarian about a vaccination program. While vaccination does not provide 100% protectio, it may prevent some abortions and it will reduce the risk of spread of infection.
- Communication with other stud farms is essential. Simple precautions will reduce the risk of disease spread to other farms and minimise disruption to horse movements usual during the stud season.
- Where there is only one case of abortion and the mare has been isolated and thorough disinfection has occurred, there is minimal risk of spread of infection and the management of the unexposed mares can continue as normal.
- In-contact pregnant mares should be isolated until they foal and for 30 days after foaling or, if another mare aborts, 30 days after the last abortion. In-contact mares include those in the same paddock as the aborted foetus and those in direct contact with the mares in the same paddock (e.g. sharing a water trough or in direct contact over a single fence line).

Felid Herpesvirus

Felid herpesvirus 1 (FeHV-1) is classified under the Order: Herpesvirales, Family: Herpesviridae, Subfamily: Alphaherpesvirinae, and genus: Varicellovirus. Characteristics of the members of the Alphaherpesvirinae are their short replication cycle, induction of lifelong latency, primary in neurons, and, in most cases, a narrow host range. Both human and animal herpesviruses are members of the Alphaherpesvirinae subfamily. Human herpes simplex viruses types 1 (HSV-1) and 2 (HSV-2), respectively, cause cold sores and genital lesions. Varicella zoster virus (VZV) is the causative agent of chickenpox, and the reactivation of latent VZV DNA causes shingles. Some of the mammalian herpesviruses, besides FeHV-1, classified under this family include bovine herpesvirus-1 (BoHV-1), which causes respiratory disease and abortions in cattle, equine herpesvirus-1 (EHV-1), which causes respiratory disease, abortions, and in some cases neurological disease in horses, Suid herpesvirus 1, also known as pseudorabies (PRV) and Aujeszky's disease virus, leading to respiratory disease, abortions, neurological disease in swine, and canid herpesvirus-1 (CaHV-1), responsible for neonatal mortality in puppies and also respiratory and ocular disease in juvenile and mature dogs. Examples of avian alphaherpesviruses are infectious laryngotracheitis virus (ILTV), causing severe respiratory disease in poultry, and Marek's disease virus (MDV), which induces immunosuppression and T-cell lymphomas.

FeHV-1 infection causes feline viral rhinotracheitis (FVR), which not only accounts for approximately half of all diagnosed feline viral upper respiratory infections, but is also an important cause of ocular lesions in cats. As is the case for other alphaherpesvirus infections, the acute phase of FVR is followed by lifelong latency. During the latent stage, viral FeHV-1 DNA persists in episomal form, primarily in the nuclei of sensory ganglion neurons. The transcription of viral RNA is very limited, and infectious virus is not produced. The reversal of the latent state, induced by natural stressors or the administration of corticosteroids, can induce viral reactivation in latently infected cells, leading to renewed production of infectious virus. Reactivated infectious virus then travels to the periphery by anterograde axonal transport, potentially leads to clinical signs (recrudescence), and can lead to viral transmission.

Since FeHV-1 is a primary pathogen of cats, with respiratory and ocular disease components that are similar to those of human herpesviruses, and latency which is easily reactivated under natural conditions, FeHV-1 infection in cats is considered to be a good natural host model to study the comparative molecular pathogenesis of acute and latent alphaherpesvirus infections and to test novel immunization strategies.

Virus Characteristics

The size of FeHV-1 virions ranges from 120 to 180 nm. They are composed of a core containing the double-stranded viral DNA genome, an icosahedral capsid surrounding the core, a tegument layer surrounding the capsid, and a lipid bilayer envelope from which glycoprotein spikes are protruding.

FeHV-1 primarily infects domestic cats, but lions and cheetahs are also susceptible. In vitro, FeHV-1 replicates only in cells of feline origin. Alphaherpesviruses that are genetically related to FeHV-1 are canid herpesvirus 1 (CaHV1) and phocid herpesviruses(PhHV) 1 and 2.

Genomic Organization

The first Sal I map of the genome of C-27 strain of FeHV-1 and determined that its size was approximately 134 kb. Grail subsequently mapped the genome of the FeHV-1 B927 strain and determined that its genome was only 126 kb in size. The genomic organization of both of these FeHV-1 strains was found to be similar to that of other varicelloviruses. Basically, the FeHV1 genome consists of two segments of unique DNA, referred to as the Unique Long (UL) and Unique Short (US) regions. The US region of the genome is flanked by a pair of identical, but inverted sequences designated the Internal Repeat Short (IRS) and Terminal Repeat Short (TRS).

The first complete genomic sequence of FeHV-1, as well as the construction and characterization of a BAC clone containing the entire viral genome. Complete genomic sequences were derived from both the FeHV-1 BAC and purified virion DNA. These data showed that the FeHV-1 genome is 135,797 bp in size and has a GC content of 45%.

A total of 78 open reading frames were predicted, encoding 74 distinct proteins. The gene arrangement was found to be colinear with that of most other varicelloviruses whose genomes have been sequenced.

All alphaherpesviruses are considered to have a replication pattern that is similar to the one of HSV-1. FeHV1 has previously been shown to contain 23 virion-associated proteins. Eight glycoproteins had initially been identified, designated as gB, gC, gD, gE, gG, gH, gI, and gL. The examination of the recently derived complete sequence showed that the FeHV-1 genome in fact contains a total of 13 envelope glycoproteins.

Most studies on the function of FeHV-1 genes have been focused on the role of envelope glycoproteins, because of their predicted role in inducing protective host immune responses and, therefore, their potential for vaccine development.

Acute Infection

FeHV-1 typically affects kittens and juvenile cats. Most kittens are protected by passive immunity until they are about 2 months of age.

The pathogenesis of FHV-1 is based upon two different mechanisms. The first is that FeHV-1 is a cytolytic virus. Examples of its cytolytic effects are ulcerations in mucosae and the cornea. The second mechanism is immune-mediated, clinically manifesting itself as stromal keratitis. An important question related to this second pathogenetic mechanism is the source of the antigenic stimulation driving this reaction.

The main sources of FeHV-1 transmission are oronasal and ocular secretions from acutely infected cats. Viral transmission can also be associated with the reactivation of latency. Kittens with residual passive immunity may not show clinical signs when exposed but become latently infected.

Following entry via the oronasal route, FeHV-1 replicates extensively in the mucosae of the upper respiratory tract and generally causes severe upper respiratory disease in susceptible animals. The incubation period varies from 2 to 6 days. The primary replication sites of FeHV-1 include the mucosae of the nasal septum, turbinate, nasopharynx, conjunctivae, and upper trachea. Replication also takes place in tonsils and mandibular lymph nodes.

Acute respiratory FeHV-1 infection is characterized initially by fever, inappetence, and sneezing, followed by serous nasal discharge, which can become mucopurulent after 5– 7 days. In addition, oral replication of the virus can result in excessive salivation and drooling of saliva. Occasionally coughing and dyspnea may occur. Oral ulceration, a typical feature of feline calicivirus infection, may occur as a result of FeHV-1 infection of the oral cavity but is uncommon.

The ocular manifestations associated with FeHV-1 infection have been reviewed by Gould. In neonatal kittens ophthalmia neonatorum has been described and can lead to serious corneal damage. Acute hyperemic conjunctivitis, leading to ocular discharge and chemosis, a feature of acute infection, occurs in association with upper respiratory signs. The formation of branched epithelial ulcers, referred to as dendritic ulceration, is a pathognomonic feature of acute ocular FeHV-1 infection. In a recent review of the etiology corneal ulcers in cats, Hartley stated "assume FHV-1 unless proven otherwise." Occasionally, larger ulcers, referred to as geographic corneal ulcers, develop. Both dendritic and geographic corneal ulceration may also result from latency reactivation. Another component of lesions associated with recrudescence is conjunctival and corneal inflammation, which is milder than seen during acute disease.

FeHV-1 is primarily an upper respiratory and ocular pathogen, with only sporadic involvement of the lungs. Viremia levels are low, thought to be related to the natural temperature sensitivity of this virus, which would favour replication in the upper respiratory tract. Exposure of pregnant queens can lead to abortion, but infection with FeHV-1 infection is not a common cause of abortion in cats. In neonatal kittens, the infection can generalize and is associated with neurological signs and a high mortality rate.

Alphaherpesvirus Latency Concepts

A hallmark of alphaherpesvirus biology is that acute infection is followed by lifelong persistence of the viral genome in latent form in nervous and lymphoid tissues. Latency and periodic reactivation of latency are integral parts of the lifecycle of alphaherpesviruses and important elements in their survival and transmission.

The latency-reactivation cycle operationally consists of three major steps: Establishment, maintenance, and reactivation.

The establishment of latency by definition requires that the virus reaches the tissue in which latency will be established. This process starts during the acute phase of viral replication at peripheral mucosal sites. Nerve endings of sensory nerves innervating viral replication sites take up viral particles and subparticles during this phase. These particles are transported within the axoplasm of the axons of these nerves by a process referred to as retrograde axonal transport. When the virus reaches the sensory ganglia, it infects neurons and other cell types. This acute infection of ganglionic cell types lasts for approximately one week. Neurons are the cell type in which latency is established. In order to accomplish this, lytic gene expression is repressed, while the latencyassociated transcript (LAT) is expressed, which yields several RNA species by splicing. These multiple species are collectively referred to as LATs. Low level or sporadic transcription of immediate-early and early genes can occur but is not sufficient to initiate a productive infection. No infectious virions can be detected in the ganglia during latent infection. The LAT RNA is spliced, and a stable intron in the form of a lariat, called the 2-kb LAT, is produced in the nucleus. The spliced LAT mRNA is transported to the cytoplasm, where several small ORFs may be translated into proteins.

During the maintenance phase of latency, the viral DNA is present in the neurons in an episomal form. The viral DNA is not totally static during the maintenance phase of latency, but transcriptional activity of the genome is limited to a region referred to as the latency-associated transcript or LAT.

The maintenance phase of latency is reversible. In other words, under the influence of certain natural or pharmacological stimuli, the reactivation of latent viral DNA can occur. Virus replication starts up again, and infectious virions then travel back to the periphery, using the same sensory nerve "highway" used to reach the ganglia. Infectious virus can be detected again by virus isolation or PCR from nasal, oral, or ocular swabs. Usually the clinical signs associated with the reactivation process are significantly milder than those seen during the primary infection, and reactivation can certainly be asymptomatic. Virus shedding resulting from reactivation is also typically at a lower level and of shorter duration than seen during primary infection. However, reactivating virus can still be a significant source of exposure and primary disease in fully susceptible hosts that are in close contact with the animal in which reactivation took place. Reactivation occurs in only a small subset of latently infected neurons, typically less than 0.05%. Latently infected neurons in which reactivation took place do not survive. This explains why sensory deficits are not associated with reactivation in sensory nerve ganglia. Since the reservoir of latently infected neurons remains large under these conditions, repeated reactivation can take place throughout the life of the host.

Our current understanding of the regulation of latency is derived primarily from studies on HSV-1 and BoHV-1.

The Role of Latency-Associated Transcripts (LATs). Acute infection of trigeminal ganglia neurons produces toxic gene expression products that make them vulnerable to damage and death. In addition, cellular DNA damage induced by viral replication stimulates the mitochondrial pathway of apoptosis. Herpesviruses try to counteract apoptosis and thus enhance their replicative ability, by encoding several antiapoptotic genes, one of which is the LAT gene. Since there is redundancy in the viral antiapoptotic capabilities during the acute phase, apoptosis of neurons during this phase is prevented fairly efficiently.

It is very important that apoptosis is prevented also during the establishment and maintenance stage of latency. This is especially crucial in permissive neurons, in which extensive viral replication has taken place during the acute phase. LAT exerts its anti-apoptotic properties through micro-RNAs (miRNAs). A mechanism by which LAT-encoded miRNA regulates apoptosis is targeting of transforming growth factor beta, a potent inducer of apoptosis.

It is important to understand the interactions between the latent viral genome and the neuron that lead to reactivation, because this is a prerequisite to ultimately controlling this process. LAT plays an important role in the in vivo reactivation of latency. In experimental studies it has been shown that spontaneous reactivation is severely impaired if the LAT gene is deleted.

The Role of Tegument Protein VP16. Thompson have recently described the central role played by the tegument protein VP16 in all phases of HSV latency. Prior to establishment of latency virus replication takes place in permissive neurons. In susceptible cells at mucosal surfaces VP16, a component of virions entering the cell, combined with cellular factors, activates the immediate early genes. Axonal transport of VP16 into neurons is inefficient, which would promote latency. In order for VP16 to initiate lytic infection, it needs to be synthesized de novo, a process which requires that neuronal inhibition be overcome.

Very interestingly, the LAT locus is considered to express riboregulators that mediate synthesis of VP16. It has been shown that, in the absence of LAT transcription, half of the neurons destined to be latently infected instead enter the lytic cycle and die. In contrast when repression is overcome, neurons become lytically infected, and the infectious virus produced spreads both within the ganglia and back to the mucosal surface where infection was initiated. The goal of lytic infection is to increase the number of latently infected cells.

Stress, leading to reactivation, is hypothesized to increase the novo production of VP16 by a mechanism that is still under investigation. The VP16 produced then initiates a feedback loop with the IE genes and results in viral reactivation in a very limited number of latently infected neurons.

The Role of Local Cell-Mediated Immune Responses. T cells, especially CD8+ T lymphocytes, have been found to be crucial for acute controlling HSV infection in sensory ganglia. Viral antigen production in trigeminal ganglia increases until 3 days after infection but is no longer detectable at 7 days after infection. As antigen production decreases, there is an increase of different types of different types of lymphoid cells, such as macrophages, natural killer cells (NK), and certain CD8+ T cells surrounding infected neurons.

It is thought that T cells, especially CD8+ T lymphocytes, inhibit reactivation from latency. Persistence of immune effector cells in trigeminal ganglia (TG) implies that low levels of viral proteins are expressed and that an immune response occurs. In a mouse HSV-1 model, it has been demonstrated that viral DNA replication, transcription, and viral protein production take place in 1 neuron per 10 TG. These individual neurons are considered to be undergoing "spontaneous molecular reactivation" and are consistently surrounded by cuffs of infiltrating white blood cells. Two mechanisms by which these infiltrating cells prevent reactivation are the production of gamma interferon and lymphocyte-mediated cytotoxicity

Reactivation of Latency

The trigeminal ganglion is considered a primary site of latency for FeHV-1 although recent studies implied other tissues as potential sites.

Spontaneous reactivation is possible but does not occur frequently. More commonly leading to the reactivation of latent FeHV-1 is the result of environmental or physiological stresses, such as changes in housing or lactation. The reactivation frequency rates have been reported to be 18% as a result of moving cats to a new environment and 40% as a result of lactation. The lag phase between the stressor leading to reactivation and the actual shedding of infectious virus is about 4–11 days, and virus excretion lasts for approximately 6 days on average. Virus excretion by cats in which a reactivation event took place ranges from 1–13 days. During this time infectious virus can be demonstrated in ocular and oronasal secretions. The reactivation can be either asymptomatic or associated with clinical signs. Symptomatic reactivation is referred to as recrudescence. Reactivation of latent viral DNA in adult cats can lead to corneal ulceration, accompanied by varying degrees of conjunctivitis. Since herpetic stromal keratitis caused by HSV-1 is the leading cause of infectious blindness in industrialized countries, ocular infection of FeHV-1 in cats is considered a very good natural host model.

The administration of corticosteroids has been reported to lead to reactivation in 70% of the latently infected cats. Infectious virus is carried by anterograde axonal transport to peripheral tissues, usually to cells at or near the site of initial infection, and is a potential source of viral transmission.

The role of reactivation in the epidemiology of alphaherpesviruses is directly related to the frequency by which it takes place. Some herpesviruses, including FeHV-1, reactivate much more easily than others from the latent state, both under natural and experimental conditions. The ease by which latent FeHV-1 DNA is reactivated is an important element in the justification of FeHV-1 infection of cats as a natural host model to study the molecular pathogenesis of herpesvirus latency and approaches to prevent it.

Diagnosis

Clinically, there is an overlap between the symptomatology of acute FeHV-1 and feline calicivirus (FCV), another major respiratory disease of cats. Distinguishing features of FeHV-1 infection are high fever and corneal ulcerations. In contrast, ulcers of the tongue, palate, and pharynx are more typical or encountered more frequently in calicivirus infections.

The most common laboratory diagnostic methods to demonstrate the presence of FeHV-1 or viral components in tissue homogenates or swabs include the direct fluorescent antibody (FA) test, virus isolation (VI), and PCR.

Fluorescent antibody testing is performed on conjunctival or corneal tissue. This test is far less commonly used now than it used to be. Topical fluorescein, used to visualize ulcers, should be avoided prior to collecting samples.

Laboratory diagnosis of acute FeHV-1 is now most commonly performed by virus isolation (VI) or PCR, using oronasal and conjunctival swab extracts as the samples. VI detects infectious virus and has been the laboratory diagnostic gold standard.

Multiple PCR assays have been described for use in the detection of FeHV-1 DNA. An excellent TaqMan-based realtime PCR assay, described by Vogtlin, targets a conserved portion of the FeHV-1 gB gene. The assay was determined to be very specific for FeHV-1, and its detection limit was between 0.6 and 6TCID50. Infectious virus titers and viral DNA correlated over a wide dilution range. The real-time PCR (qPCR) was evaluated on sequentially collected ocular fluid extracts. Early during infection, referred to as phase 1, the correlation between virus titers and qPCR signals was very high. Next, during so called phase 2, a rapid decline in infectious virus titers was seen, while the qPCR signals remained high. During the final phase, referred to as phase 3, infectious virus was no longer detectable, and the quantitative PCR signals were also declining. Analysis of the combined virus detection and qPCR results on 20 clinical samples allowed the authors to reliably define the phase of the infection during which the samples had been collected. Realizing the cost of combined testing, it was suggested to test consecutive samples by qPCR to accomplish this goal.

Maggs pointed out 3 aspects of laboratory diagnosis of FeHV-1 that can be very frustrating for the clinician. Whereas the confirmation of acute FeHV-1 is not always required, it is important to confirm that chronic lesions are caused by FeHV-1. Unfortunately, the detection of FeHV-1 or viral components in these lesions can be difficult. The second aspect of laboratory diagnosis that leads to misinterpretations is the fact that FEHV-1 or viral l DNA can be detected in samples from clinically normal cats. It was pointed out that the detection of FeHV-1 or its components can be coincidental, consequential, or causal. Differentiating between these possibilities is obviously important.

Virus neutralizing antibody titers are determined by VN tests, which are commonly used to detect prior infection or the efficacy of vaccination. Virus neutralizing antibodies can be low and slow to develop. As pointed out by Dawson, a low level of neutralizing antibodies does not imply the absence of protection against clinical disease.

Treatment and Control

Supportive Treatment: Guidelines for the management of FeHV-1-induced disease have been published by The European Advisory Board on Cat Diseases (ABCD). As is the case for many viral infections, supportive therapy is being advised. Broad spectrum antibiotics that achieve good penetration into the respiratory tract should be administered in all acute cases to prevent secondary bacterial infections. Intake of food that is palatable and flavorful is also important, since infected cats develop anorexia from the loss of their sense of smell or, less commonly, the presence of ulcers in the oral cavity. In cats with severe clinical signs, the restoration of fluids, electrolytes, and acid-base balance is required, preferably intravenously. Nasal decongestants, mucolytic drugs, and nebulization with saline can all ameliorate clinical signs. Eye drops or ointments, when used, should be administered several times a day.

Antiviral Therapy: Antiviral therapy consists of topically or systemically administered antivirals or the use of adjunctive therapies. Comparison of 8 antiviral drugs administered topically demonstrated that the highest efficacy was obtained with trifluridine, based upon its potency and corneal penetration. Second in effectiveness was idoxuridine, which has a lower cost and appears to be less irritating.

Nucleoside analogue antivirals are commonly used to treat HSV and VZV infections. They are converted into triphosphates by viral thymidine kinase and other host enzymes in infected cells and competitively inhibit viral DNA polymerase. This prevents DNA chain elongation and, as a result, disrupts viral replication.

The use of these agents against FeHV-1 infection has been largely limited to topical administration. First generation nucleoside analogues, including acyclovir and its prodrug valacyclovir, have little efficacy against FeHV-1 in vitro and moderate effect in vivo. More importantly, when administered systemically they produce serious side effects in cats, including myelosuppression, hepatotoxicity, and nephrotoxicity at therapeutic levels.

According to the guidelines of the European Advisory Board for Cat Diseases (ABCD), trifluridine is the topical treatment of choice in cats with ocular FHV-1 manifestations. Acyclovir, ganciclovir, and idoxuridine are also suggested for topical use. It was noted that, except for acyclovir, there is a lack of controlled in vivo efficacy study for these agents in the literature. The efficacy of topical application of cidofovir on primary ocular FeHV-1 infection has been demonstrated.

Although the study wasn't controlled, oral administration of famciclovir has been reported to be safe and efficacious in treating ocular signs, cutaneous disease, and rhinosinusitis induced by FeHV-1 infection.

Adjunctive therapies that are used to treat FeHV-1 infection are L-lysine, lactoferrin, and interferons. L-lysine is an antagonist of arginine; the latter has been shown to be essential for HSV-1 and FeHV-1 protein synthesis. Treatment with L-lysine, therefore, decreases viral replication and has been shown to have some inhibitory effect against both human herpesvirus and FeHV-1 infection. An issue with low dietary arginine concentrations is the pronounced susceptibility of cats to arginine deficiency.

Oral supplementation with L-lysine reduces the severity of experimentally induced FeHV-1 conjunctivitis and ocular virus shedding associated with the reactivation of latent infection. It was suggested for use early in acute disease or as a means of reducing the severity of disease and virus shedding at times of stress. It has been demonstrated that L-Lysine is safe at relatively high oral dose levels. Lactoferrin is a mammalian iron-binding glycoprotein. It has been shown to inhibit FeHV-1 replication in vitro, potentially as a result of interfering with the binding of FeHV-1 binding to its cellular receptor and viral penetration into susceptible cells.

Interferons are cytokines released by white blood cells and interfere with viral cell-to cell spread. Interferon-alpha (IFN-α) administration has been shown to decrease clinical signs associated with acute infection.

Immunity and Vaccination

Primary FeHV-1 infection induces both humoral and cellular immune responses. Active immunity induced by natural FeHV-1 infection or immunization protects cats from the disease, but not from infection. Mild clinical signs have been observed upon reexposure as soon as 150 days after the primary infection. Virus neutralizing antibody titers are generally low and in some cases undetectable after primary infection; although after further exposure to virus, they tend to rise to more moderate levels and thereafter remain reasonably stable. Since FeHV-1 targets the eye and upper respiratory tract, mucosal immune responses also play a significant role.

Passive immunity persists for 2 to 10 weeks, depending upon colostrum concentration and intake. Some kittens with low levels of maternally derived antibodies that are exposed to field virus may develop subclinical infection and latency. Alternatively, such kittens would also respond to early vaccination. Conversely, in some kittens maternally derived antibodies are high enough to still be at interfering levels at 12–14 weeks of age.

Vaccination recommendations have been provided by The European Advisory Board for Cat Diseases (ABCD) and The American Association of Feline Practitioners Feline Vaccine Advisory Panel.

The ABCD panel recommends an initial two-dose vaccination regimen: the first dose being given at 9 weeks of age and the second at 12 weeks of age. This is followed by yearly boosters.

The American Association of Feline Practitioners Feline Vaccine Advisory Panel advises that the primary immunization dose should be given as early as 6 weeks of age, with additional doses every 3 to 4 weeks until 16 weeks of age. A booster dose is to be administered 1 year following the last dose of the primary series. Subsequent booster doses are then administered every 1–3 years.

All current commercial vaccines against FVR also contain feline calicivirus (FCV) and feline panleukopenia virus (FPV) components and are collectively termed FVRCP vaccines. The protection induced by these trivalent vaccines is generally the lowest against the FeHV-1 component.

Both modified-live and inactivated FVRCP vaccines for systemic use are available in the United States. Modified-live vaccines (MLVs) are routinely used, but they have residual virulence and may induce clinical signs if administered incorrectly. Because of safety concerns, inactivated vaccines are mostly preferred for use in pregnant queens, and in cats that are infected with feline leukemia virus (FeLV) or feline immunodeficiency virus (FIV).

In addition to vaccines labeled for systemic immunization, an intranasal multivalent vaccine containing a FeHV-1 component is commercially available. Testing under experimental conditions showed that this vaccine was safe and induced protection against the clinical signs of field virus exposure within a week after vaccination, versus 2-3 weeks with a systemically administered vaccine.

New Approaches to Immunization

Virulence Genes and Deletion Mutant Vaccines, currently available vaccines cannot totally protect cats from field virus infection and, as a consequence, from field virus latency.

A better understanding of herpesvirus virulence factors is a prerequisite for the generation of safe and efficacious deletion mutant vaccines. Candidate genes for deletion are those encoding the nonessential glycoproteins gC, gE, and gG, the US3 gene encoding a protein kinase, the UL 23, gene encoding thymidine kinase. The combination of BAC cloning of herpesvirus genomes and the introduction of recombineering to rapidly generate mutants within alphaherpesviruses cloned as BACs have been very useful tools to generate mutants with vaccine potential.

Glycoprotein E (gE) is a virulence factor of FeHV-1. Glycoprotein E (gE) and glycoprotein I (gI) form a heterodimer that functions in virus cell-to-cell spread of the virus and transsynaptic spread of infection throughout the host nervous system, an important component of neurovirulence. gE/gI are nonessential glycoproteins, except for MDV. As an in vitro indicator of reduced virulence, gE/gI mutants have a smaller plaque size and reduced capacity for cell-to-cell spread. A functional gE/gI heterodimer appears to play an even greater role in the spread of VZV.

We previously constructed a gE/gI deletion mutant by conventional in vivo recombination and reported that cats vaccinated subcutaneously with high doses of the recombinant FeHV-1 strain responded with only mild clinical signs and developed strong immunity against subsequent virulent virus challenge. We also compared the intranasal and subcutaneous routes of administration of this strain and assessed its ability to induce protective immunity and prevent virus shedding after challenge. The only concern we had is that this mutant had some residual virulence when administered intranasally at high dosage levels.

Kaashoek constructed gE-, TK-, and gE-TK deletion mutants of BoHV-1 and examined their virulence and immunogenicity in calves. After intranasal inoculation, the TK mutant showed some residual virulence, whereas the gE and gE-TK mutants were completely avirulent. The calves inoculated with these deletion mutants were protected against clinical disease after challenge exposure and shed significantly less challenge virus than control calves.

Recently, an EHV-1 gE mutant was evaluated as a modified live virus (MLV) vaccine. Colostrum-deprived foals inoculated intranasally (IN) or intramuscularly (IM) with the gE mutant did not exhibit any clinical signs of respiratory disease except for mild nasal discharge in one of the IN inoculated foals on Days 1 and 3 after infection. In contrast, foals inoculated IN with the revertant had biphasic fever, mucopurulent nasal discharge, and submandibular lymph node swelling. The efficacy of the gE mutant against wild type EHV-1 challenge infection was assessed using foals previously vaccinated twice IM with 10^5 or 10^6 plaque-forming units (pfu) of the gE-mutant at an interval of 3 weeks. These foals exhibited no respiratory disease signs after IM immunization and developed a good virus neutralizing antibody response to EHV-1 after the second dose. Following a wildtype EHV-1 challenge infection, vaccinated foals showed milder clinical symptoms than foals vaccinated with a placebo, and challenge virus shedding was significantly reduced.

The thymidine kinase (TK) gene of alphaherpesviruses is a virulence factor. Comparisons of the amino acid sequences of herpesvirus TK proteins showed that these proteins are highly divergent, sharing only short regions of imperfect amino acid identity. Nunberg first identified the TK gene of FeHV-1 using PCR with highly degenerate oligonucleotide primers. Yokoyama inserted the gene encoding the feline calicivirus capsid protein into the TK locus of FHV-1 and designated the recombinant C730ldfTKCap. In a pilot study, 2 cats were inoculated intranasally and orally with C730ldfTK-Cap, and one cat was inoculated via the same routes with C730ldfTK. Virus-neutralizing (VN) antibody against both FeHV-1 and FCV was induced with C730ldfTK-Cap, and against FeHV-1 with C730ldfTK.

The US3 gene of FeHV-1 encodes a serine/threonine protein kinase (PK), and its amino acid sequence is conserved in the subfamily Alphaherpesvirinae. Possible functions of PK include blocking of apoptosis induced by both viral and cellular proteins, regulation of the nuclear egress of progeny nucleocapsids, and control of the morphology of infected cells. Kimman demonstrated that a PK-mutant of pseudorabies virus (PRV) has strongly reduced virulence, and animals inoculated with PK-gE-PRV mutant and subsequently challenged with wildtype virus has reduced virus shedding.

Glycoprotein C (gC) homologues have been extensively studied in several alphaherpesviruses. gC homologues are nonessential for herpesvirus replication in vitro, but they mediate several important biological functions. First of all, gC is involved in the initial step of viral attachment by interacting with heparan sulfate on cell surface, as demonstrated in HSV-1, PRV, BHV-1, and EHV-1. gC deficient mutants attach to cells with reduced efficiency. Secondly, gCs of HSV-1 and -2 can bind the complement component C3b. Binding of this complement factor may protect herpesvirus-infected cells from complementmediated lysis. Viruses lacking complement-binding domains are less virulent than wild-type virus. The gC of FeHV-1 has been shown to be the dominant heparin-binding glycoprotein that mediates the initial stage of viral adsorption, as observed in other herpesviruses. However, it remains to be determined whether FeHV-1 gC protects virus-infected cells from complement-mediated lysis.

Willemse first determined a partial sequence of gC. They also found that the adjacent UL45 gene can be cotranscribed with gC. The complete sequence of FeHV-1 gC was later determined by Maeda. Based on the amino acid sequence deduced from the nucleotide sequence, they predicted that gC is a membrane glycoprotein containing a characteristic N-terminal hydrophobic signal sequence, nine potential N-linked glycosylation sites, and C-terminal transmembrane and cytoplasmic domains. Maeda further demonstrated that gC is the major heparin-binding glycoprotein involved in the initial step in virus adsorption to cells as observed in gCs of other herpesviruses. In addition, they found that gC can agglutinate murine red blood cells, and that infection of FeHV-1 is inhibited by the addition of soluble heparin in cells cultures.

The gG glycoprotein of herpesviruses interacts with chemokines, which are involved in the regulation of leukocyte trafficking and function and the regulation of inflammation and immunosurveillance. The gG glycoprotein of alphaherpesviruses can exist in three different forms: Membranebound full length, membrane bound truncated or secreted. The full length form, present in FeHV-1 and EHV-1, can also exist as a truncated secreted form. The secreted form functions as a viral chemokine-binding protein (vCKBP) and is now classified under the vCKBP-4 subfamily.

Van de Walle used EHV-1 as a model to provide the first molecular determination of the residues in gG of EHV-1 involved in chemokine binding and interaction with target cells. In a very recent study, Thormann constructed recombinant viruses to show that the ability of the gG of EHV-1 to interfere with chemokine is not entirely mediated by its chemokine-binding region.

The gG of FeHV-1 exists in booth membrane-bound and secreted forms. The secreted form shows in vitro binding to bind to a number of chemokines. The membrane bound displays true viroreceptor characteristics.

Virulence characteristics of the gG of several alphaherpesviruses have been investigated. It has been shown previously that the deletion of gG in PRV does not have a significant effect on viral virulence. In contrast, the administration of a gG-deleted ILTV to birds, the natural host of this virus, showed that the gG deletion resulted in significant reduction in virulence. Importantly, virulence could be restored with a revertant, and the transcription of genes adjacent to the gG deletion was not affected by the gG deletion. Immunization with the gG deletion mutant was shown to be protective against virulent virus challenge in experimental birds.

Herpesviruses have multiple immune evasion genes with various evasion mechanisms.

UL49.5 is a gene present in the genome of several members of the varicellovirus genus, such as EHV-1, BoHV-1, and PRV. UL49.5 inhibits transporter associated with antigen processing (TAP) and downregulates cell-surface expression of major histocompatibility complex (MHC) class I molecules.

A BoHV-1 UL49.5 null mutant was shown to no longer have the TAP inhibition and MHC-I downregulation properties of the parent virus. In a follow-up study, the pathogenicity and immune responses in calves infected with BoHV-1 UL49.5 null mutant and the parent wild type strain were compared. Both strains replicated similarly in the nasal epithelium, and both groups had similar clinical scores. BoHV-1 antigen-specific CD8+ T-cell proliferation as well as CD8+ T-cell cytotoxicity in calves infected with the BoHV-1 UL49.5 null mutant peaked by 7 days after infection, 1 week earlier than in calves infected with the wild type strain. In addition, virus neutralizing antibody (VN) titers and IFNγ-producing peripheral blood mononuclear cells (PBMCs) in the UL49.5 mutant virus-infected calves also peaked 7 days and 14 days earlier, respectively. This study indicated that while immune responses peak earlier, deleting UL49.5 by itself did not sufficiently attenuate this alphaherpesvirus to make it a vaccine candidate.

Generating Mutants by BAC Clone Recombineering. BAC cloning and recombineering are two state-of-the-art techniques to facilitate the process of mutagenesis. BACs are single copy F-factor-based plasmid vectors, which can stably hold 300 kb or more of foreign DNA. The BACs' larger capacity and greater stability over the other vectors have enabled the cloning of an entire herpesvirus genome into a single plasmid. These properties have also made BAC the vector of choice for the cloning of herpesvirus genomes.

Recombineering is a powerful method for fast and efficient manipulation of the BAC. It allows DNA cloned in E. coli to be modified via lambda (λ) red-mediated homologous recombination, obviating the need for restriction enzymes and DNA ligases. Specific bacterial strains, for example, E. coli SW105, have been constructed for this purpose. A defective λ prophage (mini-λ) is inserted into the E. coli genome and encodes heat-shock inducible genes that make recombineering possible. Linear DNA (PCR product, oligonucleotide, etc.,) with sufficient homology in the 5 and 3 ends to a target DNA molecule already present in the bacteria (plasmid, BAC, or the bacterial genome itself) can be electroporated into heat-shocked and electrocompetent bacteria cells and undergoes homologous recombination with the target molecule. Utilizing recombineering techniques, site-specific mutations can be introduced anywhere in the viral genome. All mutagenesis steps can be strictly controlled and analyzed in E. coli, and the manipulated viral genome can be stably maintained in the E. coli.

The entire FeHV-1 genome was previously cloned as a BAC in our lab, from which the complete FeHV-1 genomic sequence was derived . The BAC-cloned virus was characterized in vitro and in vivo. Prior to defining the in vitro growth characteristics of the BAC-cloned virus, the BAC cassette was excised from the cloned virus genome. We then performed plaque size analysis and constructed multiplestep growth curves for the FeHV-1ΔBAC and its C-27 parent strain. Plaques produced by the C-27 strain and FeHV1ΔBAC virus were morphologically undistinguishable from each other. The mean plaque diameter of the FeHV-1ΔBAC virus was 101.05% of that of the C-27 parent strain and not significantly different. Multistep growth curve analysis showed that they can grow to a similar titer.

To investigate possible attenuation resulting from BAC cloning itself, a preliminary challenge experiment was carried out, using four specific-pathogen-free (SPF) cats. Two cats were inoculated intranasally with the FeHV-1ΔBAC virus, and the other two cats were inoculated intranasally with either the C-27 strain or cell culture medium. The main conclusion from the in vivo experiment was that the BAC clone-derived virus behaved very similarly to its C-27 parent strain both in vitro and in vivo, making it an excellent starting platform for introducing mutations aimed at deleting virulence-inducing genes from the FeHV-1 genome.

Mucosal Vaccination and Epitope-based Vaccines

A major goal of strategies to immunize against alphaherpesvirus infections is to prevent primary infection, which would in turn prevent primary disease and the establishment of latency and subsequent latency reactivation. Latency reactivation has been shown to occur frequently, leading to virus shedding, which is asymptomatic in most cases. Natural infection provides protection against reinfection of primary mucosal replication sites for a certain period of time. This provides a rationale for the development of immunization strategies at the mucosal level.

Innate immune responses are the first to develop after natural infection. The recognition of alphaherpesvirus components by toll-like receptors is an important mechanism for induction of these responses. HSV and its components bind to TLR 2, 3, 7, and 9. Synthetic agonists have been designed to transiently activate the innate immune response.

In human medicine, the majority of the efforts to develop immunization strategies against herpesvirus infections have been focused on the prevention of genital herpes. However, it is also well recognized that ocular HSV-1 infection is a leading worldwide cause of herpetic keratitis, which can lead to corneal blindness. Like is the case for FeHV-1, the most severe ocular HSV-1 infections are the result of repeated reactivation events. It is clear that mucosal delivery is the best approach to generate secretory immunity and cytotoxic Tcell responses at mucosal sites.

Long-term efforts to immunize against human alphaherpesvirus infections have included subunit vaccines, modified-live vaccines, replication-defective vaccines, viral vector vaccines, and naked DNA vaccines. Despite these efforts, there are no licensed vaccines available.

One of the current approaches to mucosal immunization focuses on the development of a multiepitope self-adjuvant lipopeptide vaccine. A recent study of this approach by the group that has pioneered it highlights its promise, but also the hurdles that still have to be overcome. They point out that, based upon recent trials, the induction of neutralizing antibodies is not sufficient for protection. Implied from these results is that the induction of appropriate and adequate protective T-cell responses is a crucial part of the development of protective immunity. The essential components of a protective immune response can be the prevention of primary infection or the prevention or reduction of reactivation events. It is clear from their work, and that of others, that individuals that are latently infected with HSV, have frequent reactivation events associated with virus shedding. This reactivation is not associated with clinical signs in most individuals, which are therefore termed asymptomatic individuals. In contrast, the term symptomatic individuals is used for those in which frequent reactivation is associated with clinical signs. An important element of the strategy is to characterize the unique T-cell repertoire in HSV-positive individuals that do not suffer from frequent symptomatic reactivation. It has been determined that a set of human Tcell epitopes from HSV-1 gB and gD are strongly recognized by T-cells from asymptomatic individuals, but not by T cells from symptomatic individuals. In contrast, another nonoverlapping set of gB and gD epitopes is recognized by symptomatic individuals. The results of recent immunization of asymptomatic HLA transgenic rabbits showed that immunization with asymptomatic CD8+ epitopes from HSV-1 gD induced strong CD8+ immune responses and reduced HSV-1 shedding and tears and corneal lesions following ocular challenge virus administration.

The following five existing hurdles need to be overcome: (1) reasons for suboptimal immunity resulting from natural infection, (2) optimal effector mechanisms for protective immunity against the acute and latent phases of the disease, (3) knowledge about immunoevasive strategies, (4) distinction between protective versus pathogenic antigens, and (5) design of a an appropriate vaccine delivery system. They recognize that candidate vaccines need to be tested in relevant animal models if they cannot be directly evaluated in the natural host. An already existing human HLA transgenic rabbit model and the development of a similar guinea pig model are crucial tools in this respect.

In Vitro Approaches to Molecular Pathogenesis

Ocular infection with FeHV-1 results from viral exposure of conjunctival and corneal tissue. Since corneal lesions are an important disease manifestation, both during both the acute phase and as a result of reactivation, finding an effective therapy against development of ocular disease has high priority. Sandmeyer have reported the development of primary corneal cell culture system, which is useful for in vitro pathogenesis of ocular disease and also for the testing of potential antivirals. Using this system they showed that IFN-α was not toxic to ocular cells and had a limited effect of virus production in FeHV-1-infected corneal cells. They speculated that a combination of IFN-α and other antivirals may act synergistically.

Pathogenesis studies of FeHV-1 have almost exclusively been done on live animals. Since the tracheal mucosa is an important replication site of FeHV-1, tracheal organ cultures are a good in vitro model to study viral invasiveness and local immune responses. Leeming established feline tracheal organ cultures and showed that these could be maintained for at least 5 days. Infection of these cultures at different multiplicities of infection (MOI), ranging from 0.1 to 100, showed that the virus replicated extensively in these cultures and produced coalescing necrosis of tracheal epithelium and disruption of ciliary activity.

Since mucosal surfaces are the primary replication sites of FeHV-1, it is important to understand viral replication strategies and the local immune responses generated at these sites to better combat this mucosal pathogen. As indicated earlier, it is well known that systemically administered vaccines can prevent clinical signs but cannot prevent reinfection and the associated development of latency.

Quintana recently developed an equine respiratory epithelium cell culture system consisting of culturing dissociated primary epithelial cells at a liquid air interface. This is a meaningful in vitro system since epithelial cells not only provide a physical barrier against viral invasion, but also play a significant role in development of immunity by expressing toll-like receptors, by secreting cytokines, chemokines, and host defense peptides and by playing some role in antigen presentation. It was shown that epithelial cell cultures grown under these conditions were morphologically similar to intact airway epithelium. These cultures were also shown to be immunologically competent, but some properties were altered by in vitro culture under sterile conditions. The authors concluded that the addition of antigenic stimuli and immune cells could reverse this situation.

Mucosal explants have recently been shown for several herpesviruses to be an excellent system to study kinetics of viral invasion, as determined by the ability of a particular herpesvirus to get across the epithelial basement membrane. This system has been used to compare the invasiveness of different herpesviruses. It can, however, also be used for strain comparison and to study the role of individual or combinations of viral genes as determinants of viral virulence.

FeHV-1 infection of cats is an excellent natural host model to study mechanisms involved in establishment, maintenance, and reactivation of latency., latency is established in all cats following natural infection and is readily reactivated by a variety of natural stimuli or administration of corticosteroids.

De Regge reported the development of a homologous in vitro model to study the interaction of alphaherpesviruses and trigeminal ganglion neurons. The system consists of two concentric culture chambers. The inner and outer chambers are separated from one another by a silicon barrier, which is impermeable to both virus and cell culture medium. After 2-3 weeks in culture, axons from neuronal cell bodies present in the inner chamber grow through the silicon barrier into the outer chamber. Infection of these axons, either with HSV-1 or PRV, exclusively led to infection of neurons in the inner chamber and the subsequent spread of infection from these neurons to other neurons and nonneuronal cells. This system thus allows an in vivo-like infection of neuronal cells via retrograde axonal transport. It is, therefore, very useful to study mechanisms involved in latency establishment, maintenance, and reactivation. In a followup study, De Regge used this system to examine the role of IFN-α, an important component of the innate immune system. The data showed that IFN-α was indeed able to establish latency in these cultures and that latency was maintained after its removal. LAT transcripts, a prominent feature of latency, were detected in the cultures by RT-PCR and the latent viral DNA could be reactivated by treatment with forskolin.

A New Approach to Antiviral Therapy

Ribonucleic acid interference (RNAi), initiated by chemically synthesized 21-mer or 27-mer small interfering RNAs (siRNA), is an alternative method to the use of standard antiviral therapy. Wilkes and Kania have explored the potential of this method in vitro. The initial target was gDspecific mRNA, based upon the fact that the gD glycoprotein plays an important role in viral attachment to susceptible cells and also in the induction of protective neutralizing antibody responses. Two of the six siRNAs they tested induced a significant reduction of virus replication in CRFK cells infected with FeHV-1. In a follow-up study Wilkes and Kania selected siRNAs specific for the FeHV-1 DNA polymerase mRNA, the gD mRNA, or a combination of both. The hypothesis behind the targeting of the DNA polymerase was that more complete inhibition of viral replication would occur when an early rather than a late transcript was targeted. This proved to be the case, since the highest level of inhibition was obtained with a combination of 2 siRNAs targeting the FeHV-1 DNA polymerase transcript. Potential in vivo use of this approach is based upon the fact that siRNAs can be taken up effectively when applied to mucosal surfaces.

Suid Herpesvirus

Species Suid herpesvirus 1 (SuHV-1, pseudorabies virus, Aujeszky's disease virus) is in the family Herpesviridae, subfamily Alphaherpesvirinae, genus Varicellovirus. SuHV-1 is capable of causing reduced fertility in boars and sows, fetal resorption, mummification, abortion, and stillbirth in sows, and fatal meningoencephalitis in young pigs, sheep, cattle, and dogs. Strains of SuHV-1 vary in their ability to invade the placenta and fetus; however, abortion without invasion of the conceptus may occur secondary to fever and nervous disease in the sow. When the virus does invade the placenta and fetus, characteristic gross and microscopic lesions are produced. There is a strong correlation between the ability of strains to cause syncytial formation in cell culture and their ability to produce disease in pigs and cattle.

The disease has an almost worldwide distribution, and the carrier sow is the most important source of the virus. Recrudescence and shedding through nasal mucus may occur following stress, and there is some evidence for long-distance airborne transmission, although this would appear not to be a major threat. Boars may carry SuHV-1, spreading it through nasal mucus and semen. The virus can survive in a pen for up to 2 weeks.

The virus enters the nasal passages of the susceptible pig and passes via the pharyngeal and olfactory nerves to the brain, where it rapidly spreads to the rest of the brain, causing nonsuppurative meningoencephalitis. It has affinity for the respiratory tract, and may induce severe rhinitis, tonsillitis, and pneumonia.

Abortion usually occurs about 10 days after the onset of clinical illness in the sow. If the placenta and fetus are invaded, lesions consist of multifocal coagulative necrosis of chorionic villi and focal coagulative necrosis in many organs, including liver, adrenal, and spleen. Intranuclear inclusions have been observed in trophoblast and interstitial cells and hepatocytes around areas of necrosis.

Virus may be demonstrated in nasal or vaginal swabs and fetal tissues by isolation and fluorescent antibody or ELISAs; PCR assays are also available.

Although suid herpesvirus 1 has been eradicated from domestic swine in numerous countries, this virus remains enzootic in wild and domestic swine in many parts of the world, causing substantial adverse economic impact to swine production in the countries where it occurs. Swine are the primary host and reservoir for the virus, which causes a uniformly fatal disease when transmitted to a wide variety of nondefinitive hosts. Virus is shed in the saliva and nasal discharges of swine, so that transmission can occur by licking, biting, and aerosols. Virus is not shed in significant titers in the urine or feces. The contamination of livestock feed or the ingestion of infected carcasses by swine is common, and ingestion of virus-contaminated material, including pork, is probably the most common source of infection for carnivore hosts. Rats may contribute to farm-to-farm transfer, and sick or dead rats and other feral animals are probably the source of infection for dogs and cats. Direct transmission from swine to cattle by aerosol has been described.

Some swine that have recovered from pseudorabies may shed virus continuously in their nasal secretions. Others from which virus cannot be isolated by conventional means may yield virus from explant cultures derived from the tonsil. Pseudorabies virus DNA can be demonstrated in the trigeminal ganglia of recovered swine, but there is debate about the relative significance of lymphoreticular cells and nerve cells as sites for latency.

Clinical Signs in Swine

In herds in which the disease is enzootic, reactivation of virus occurs without obvious clinical signs, but the spread of the virus within a susceptible (nonimmune) herd may be rapid, with the consequences of primary infection being influenced markedly by age and, in sows, by pregnancy. Pruritus, which is such a dominant feature of the disease in secondary hosts such as cattle, is rare in swine. Importantly, in the absence of vaccination in virus-free countries, the eradication of pseudorabies virus from domestic swine provides a fully susceptible domestic swine population and heightens the need for biosecurity.

Pregnant sows: In fully susceptible herds, up to 50% of pregnant sows may abort over a short period of time, as a result of rapid spread of infection from an index case or carrier. Infection of a sow before the 30th day of gestation results in death and resorption of embryos (embryonic loss), whereas infection after that time can result in abortion. Infection in late pregnancy may terminate with the delivery of a mixture of mummified, macerated, stillborn, weak, and normal swine, and some of these pregnancies may be prolonged. Up to 20% of aborting sows are infertile on the first subsequent breeding, but do eventually conceive.

Piglets: Mortality rates among piglets born to nonimmune dams depend somewhat on their age, but approach 100%. Maternal antibody is protective, and disease in piglets born to recovered or vaccinated sows is greatly diminished in severity, with recovery the usual outcome.

Weaned, growing, and mature swine: The incubation period is typically about 30 h. In younger pigs, the course is perhaps 8 days, but it may be as short as 4 days. Initial signs include sneezing, coughing, and moderate fever (40 °C), which increases up to 42 °C in the ensuing 48 h. There is constipation during the fever; the feces are hard and dry, and vomiting may occur. Pigs are listless, depressed, and tend to remain recumbent. By the 5th day there is incoordination and pronounced muscle spasm, circling, and intermittent convulsions accompanied by excess salivation. By the 6th day, swine become moribund and die within 12 h. In mature swine the mortality rate is low, usually less than 2%, but there may be significant weight loss and poor growth rates after recovery.

Clinical Signs in Nondefinitive Hosts

Important secondary hosts include cattle ("mad itch"), dogs ("pseudorabies"), and cats. Disease in secondary hosts is sporadic and occurs where there is direct or indirect contact with swine. Infection is usually by ingestion, less commonly inhalation, and possibly via minor wounds. In cattle the dominant clinical sign is intense pruritus. Particular sites, often on the flanks or hind limbs, are licked incessantly; there is gnawing and rubbing such that the area becomes abraded. Cattle may become frenzied. There is progressive involvement of the central nervous system; following the first signs, the course leading to death may be as short as a few hours, and is never longer than 6 days.

In dogs, the frenzy associated with intense pruritus and paralysis of the jaws and pharynx, accompanied by drooling of saliva and plaintive howling, simulates true rabies; however, there is no tendency for dogs to attack other animals. In cats, the disease may progress so rapidly that frenzy is not observed.

Pathogenesis and Pathology

After primary oral or intranasal infection of swine, virus replicates in the oropharynx. There is no viremia during the first 24 h and it is difficult to identify virus at any time. However, within 24 h, virus can be isolated from various cranial nerve ganglia and the medulla and pons, to which virions have traveled via the axoplasm of the cranial nerves. Virus continues to spread within the central nervous system; there is ganglioneuritis at many sites, including those controlling vital functions.

The relative lack of gross lesions even in young swine is notable. Tonsillitis, pharyngitis, tracheitis, rhinitis, and esophagitis occasionally may be evident, with formation of a diphtheritic pseudomembrane overlying the affected mucosa. Similarly, discrete small white or yellow foci of necrosis may sometimes be present in the liver and spleen. Microscopically, the principal findings in both swine and secondary hosts are in the central nervous system. There is a diffuse nonsuppurative (predominantly lymphocytic) meningoencephalitis and ganglioneuritis, marked perivascular cuffing, and focal gliosis associated with extensive necrosis of neuronal and glial cells. Typical intranuclear herpesvirus inclusions are uncommon in the lesions in affected swine.

Diagnosis

The history and clinical signs often suggest the diagnosis, which is confirmed by histopathology and virus detection methods. Immunohistochemistry or fluorescent antibody staining of frozen tissue sections, PCR assay, virus isolation or serum neutralization assay are used for confirmation. Enzyme immunoassay has been approved as a standard test in several countries and is used in association with vaccination and eradication programs.

Immunity, Prevention and Control

Management practices influence epidemiologic patterns of suid herpesvirus 1 infection and disease in swine. Losses from severe disease occur when susceptible pregnant sows or swine less than 3 months old, born to nonimmune sows, are infected. Such a pattern is likely to occur when the virus is newly introduced into a herd or unit within a farm. When breeding sows are immune with adequate antibody levels, overt disease in their progeny is not observed or is reduced greatly. Where breeding and growing/finishing operations are conducted separately, significant losses from pseudorabies occur when weaned swine from several sources are brought together in the growing/finishing unit, but the disease in these older swine is less severe than that in piglets. If care is taken to prevent the entry of virus, the move toward complete integration of swine husbandry (so-called farrow-to-finish) operations provides an ideal situation by which to produce and maintain pseudorabies-free herds and thus avoid the cost of disease and the problems inherent to vaccination.

Vaccination of swine in areas where the virus is enzootic can reduce losses. Recombinant DNA, deletion-mutant, live-attenuated, and inactivated vaccines are all available commercially, but they do not prevent infection or the establishment of latent infection by the wild-type virus. A pseudorabies vaccine from which both the thymidine kinase and a glycoprotein gene have been deleted, and the E1 gene of classical swine fever (hog cholera) virus inserted, provides protection against both pseudorabies and classical swine fever in regions where both viruses are enzootic. Vaccination of secondary hosts is rarely undertaken, because of the sporadic incidence of the disease.

Duck Virus Enteritis

Duck virus enteritis (DVE) is an acute, sometimes chronic, contagious virus infection that occurs naturally only in ducks, geese and swans, all members of the family Anatidae of the order Anseriformes. The disease is a potential threat to commercially reared, domestic and wild waterfowl. The aetiological agent, Anatid alphaherpesvirus-1 or DVE virus (DVEV), is a member of the Alphaherpesvirinae subfamily of the Herpesviridae, genus Mardivirus. DVE may also be referred to as duck plague, anatid herpes, eendenpest, entenpest and peste du canard. The infection has not been reported in other avian species, mammals or humans.

In domestic ducks and ducklings, DVE has been reported in birds ranging from 7 days of age to mature breeders. In susceptible flocks the first signs are often sudden, high and persistent mortality with a significant drop in egg production in laying flocks. In domestic ducks the incubation period ranges from 3 to 7 days. Mortality usually occurs 1–5 days after the onset of clinical signs and is often more severe in susceptible adult breeder ducks. In chronically infected partially immune flocks only occasional deaths occur. Recovered birds may be latently infected carriers and may shed the virus in the faeces or on the surface of eggs over a period of years. DVE limited solely to Muscovy ducks has been observed in the USA.

Clinical signs and gross pathology associated with a DVE outbreak vary with the species, immune status, age and sex of the affected birds, and with the virulence of the virus. Similarly, as infection progresses within a flock, more clinical signs are typically observed. In breeder ducks the range of signs include sudden deaths, photophobia associated with partially closed, pasted eye-lids, polydipsia, loss of appetite, ataxia, and nasal discharge. Birds often have ruffled feathers, watery diarrhoea and soiled vents. Sick birds may maintain an upright stance by using their wings for support, but their overall appearance is one of weakness and depression. In ducklings 2–7 weeks of age, losses may be lower than in older birds, and the signs associated with DVE include dehydration, loss of weight, conjunctivitis and serous ocular discharge, a blue colouration of the beaks and blood-stained vents.

At necropsy, adult ducks that have died are typically in good body condition. In mature males, prolapse of the penis may occur. In mature females, haemorrhages may be observed in ovarian follicles. The gross lesions are characterised by vascular damage, with tissue haemorrhages, free blood in the body cavities and intestinal lumen and a range of lesions affecting the digestive tract mucosa. Specific digestive mucosal lesions may be found in the oral cavity, oesophagus, caeca, rectum and cloaca.

Lesions undergo alterations as the disease progresses, from initial macular surface haemorrhages, to yellow-white crusty plaques, then green superficial scabs. Lesions may coalesce and be covered with a diphtheritic membrane. These latter lesions progress with the course of the disease and include initial mucosal haemorrhages and eruptions and intense annular congestion, leading to pseudo-membranous or diphtheritic mucosal lesions. Necrotic degenerative changes are evident in the lymphoid and parenchymatous organs. In the liver this manifests as irregularly distributed pinpoint haemorrhages and white foci giving a speckled appearance. In ducklings lesions of the lymphoid tissues tend to be more prominent than visceral haemorrhages. Collectively, these lesions are pathognomonic for DVE. The pathology and histopathology of DVE in white Pekin ducks has been reviewed. Microscopic lesions are characterised by vascular damage and its consequences in visceral organs. Eosinophilic intranuclear inclusions and cytoplasmic inclusions in epithelial cells of the digestive tract are typically present. Birds that recover from natural infection are suggested to be immune to re-infection, but latency (in the trigeminal ganglion) and reactivation of virus is recognised.

Diagnostic Techniques

Table: Test methods available for the diagnosis of duck virus enteritis and their purpose.

Method	Purpose					
	Population freedom from infection.	Individual animal freedom from infection prior to movement.	Contribute to eradication policies.	Confirmation of clinical cases.	Prevalence of infection – surveillance.	Immune status in individual animals or populations postvaccination.
Agent identification						
Virus isolation	+	–	++	+++	+	–
Antigen detection	+	+	++	++	+	–
Real-time PCR,	++	++	++	+++	++	–
Conventional PCR,	++	++	++	+++	++	–
	++	++	++	+	++	–
Detection of immune response						
Microtitre plate VN	+++	+++	+++	+	+++	+++
VN in duck embryos	+++	+++	+++	++	+++	+++
ELISA	+	+	+	+	++	+

Key: +++ = recommended method, validated for the purpose shown; ++ = suitable method but may need further validation.

+ = may be used in some situations, but cost, reliability, or other factors severely limits its application.

– = not appropriate for this purpose.

n/a = purpose not applicable.

PCR = polymerase chain reaction.

LAMP = loop mediated isothermal amplification.

ELISA = enzyme-linked immunosorbent assay.

VN = Virus neutralisation.

Identification of the Agent

Primary isolation of the virus is best achieved from samples of liver, spleen or kidney tissue, which have been homogenised in buffered saline containing antibiotics and clarified by low-speed centrifugation (1800 g). Isolation may be attempted by inoculating such homogenates into cell cultures or duck embryos.

Cell Cultures

Cell culture is reported as the method of choice for isolation of DVEV, but may not always be successful. If attempted, isolations may be made in primary duck embryo fibroblasts (DEF) or, preferably primary Muscovy duck embryo fibroblasts (MDEF). Muscovy duck embryo liver (MDEL) cells are thought to be even more sensitive. Cell monolayers grown in Eagle's minimal essential medium (MEM) containing 10% fetal calf serum (FCS), 2 mM glutamine, and 0.17% sodium bicarbonate and gentamicin are washed with serum-free MEM and then inoculated with the clarified sample homogenate suspected to contain DVEV. After incubation for 1 hour at 37 °C to allow for virus adsorption, the cultures are maintained on MEM containing 2% FCS, 2 mM glutamine, and 0.17% sodium bicarbonate and gentamicin, and incubated in an atmosphere containing 5% CO_2. The cytopathic effect (CPE) is characterised by the appearance of rounded clumped cells that enlarge and become necrotic 2–4 days later. Cultures should be stained with a labelled antibody conjugate using a direct or indirect method to identify the virus. Cells can also be fixed and then stained with haematoxylin and eosin to show syncytial formation, intranuclear inclusions and marked cytoplasmic granulation. It has been reported that the isolation of DVEV in MDEF cells is favoured by incubation at temperatures between 39.5 °C and 41.5 °C. However, an elevated temperature does not appear to be essential for isolation, which is often carried out at 37 °C. More than one passage in cell culture may be necessary to isolate the virus.

This virus isolation method in cell cultures may be modified to a plaque assay by overlaying the cell monolayer with maintenance medium containing 1% agarose. As the virus can be cell associated, sequential passaging should be carried out by trypsinising potentially infected cells and replanting them, as well as inoculating fresh cells with infected culture supernatant from the previous passage.

Duck Embryos

Primary virus isolations can be made by inoculation on to the chorioallantoic membrane (CAM) of 9– 14 day embryonated Muscovy duck eggs. The inoculated embryos should be monitored daily and those that die within 72 hours should be removed. Virus may be harvested from embryos that survive for 72–120 hours. Before harvest, inoculated embryonated eggs should be chilled at 4 °C for 4 hours or overnight to kill the embryos before further manipulations. The embryos may die, showing characteristic extensive haemorrhages 4–10 days after inoculation. Two to four serial blind passages of the homogenised CAMs may be necessary before isolation can be effected.

Embryonated chicken eggs are not very susceptible to infection with field strains of DVEV. The virus can nevertheless be adapted to chicken embryos by serial passages. Pekin duck embryos vary in their susceptibility to strains of DVEV.

Immunological Methods

Serological methods used to confirm the identity of newly isolated virus include neutralisation assays performed in either embryonated eggs or cell cultures. A plaque assay for DVEV in duck embryo cell cultures has been described. A microtitre assay using primary MDEF or MDEL cells can be used. Provided a hyperimmune antiserum of sufficiently high titre is used, a fluorescent antibody test (direct or indirect) for DVEV in DEF, MDEF or MDEL cells is the next most sensitive assay after isolation in 1–9-day old ducklings. A reverse passive haemagglutination test for DVEV has been described, but it is reported to be less sensitive than immunofluorescence and plaque assays. An avidin–biotin–peroxidase method of immunoperoxidase staining to detect DVEV antigen in formalin-fixed, paraffin-embedded sections of liver and spleen from experimentally infected birds has been described. The identity of the virus may also be confirmed by negative stain electron microscopy, but this alone is not positive confirmation that the herpesvirus observed is DVEV. Immunoelectron microscopy has been developed recently using DVE hyperimmune serum.

Nucleic Acid Recognition Methods Polymerase Chain Reaction

Several protocols for the detection of DVEV by conventional polymerase chain reaction (PCR) have been reported. Real-time quantitative PCR protocols for the detection of DVEV have also been reported.

A loop-mediated isothermal amplification (LAMP)-based method for the detection of DVEV DNA has been published. Primers have been identified that are able to amplify DNA from DVEV present in various tissues, including oesophagus, liver and spleen, from an original outbreak and after passage from Muscovy duck embryos. Tissues are preferred to cloacal swabs as DVEV shedding may be intermittent in infected waterfowl.

The following is an example protocol for conventional PCR for detection of DVEV; other protocols exist.

Extraction of Viral DNA

This DNA extraction procedure can be used on disrupted cell suspensions from DVEV-infected tissue culture, 10% ground tissue suspensions, or cloacal swab material in transport medium. This method is used to prepare DVEV DNA for the known positive PCR controls.

All product transfers in steps i to v are performed in a biological safety cabinet.

- For a 10% ground tissue suspension, add 400 µl to a 1.5 ml microfuge tube and microfuge at 16,000 g for 5 minutes. Transfer the supernatant to a new tube.
- For tissue culture suspensions and cloacal swab material, add 400 µl of the sample, or supernatant from step i above, to a 1.5 ml tube and microfuge at 16,000–20,000 g for 45 minutes to pellet the virus.
- Discard the supernatant and resuspend the pellet with 200 µl of Tris/ethylene diamine tetra-acetic acid (EDTA) buffer (10 mM Tris/HCl, pH 8.0, 1 mM EDTA).
- Add 10 µl of a 5 µg/µl proteinase K solution to give a final concentration of 0.2 µg/µl, mix thoroughly, and incubate at 56 °C for 1 hour.
- Add 25 µl of 10% sodium dodecyl sulfate (SDS) solution to give a final SDS concentration of 1%, mix thoroughly, and incubate at 37 °C for 1 hour.
- Add 15 µl of 5 M NaCl to give a final concentration of 0.3 M and mix thoroughly.
- Add 300 µl of fresh phenol buffered with Tris/HCl, pH 8.0, to the tube, and mix by inverting 50 times.
- Microfuge the tube at 16,000 g for 5 minutes and transfer the top aqueous phase (sample) to a new tube. ix) Repeat the phenol extraction steps vii and viii once more.
- Add 500 µl of ether to the tube, mix thoroughly, and microfuge at 16,000 g for 1 minute.

- Discard the top aqueous phase (ether) and repeat the ether extraction step once more. Heat the tube with the lid open at 56 °C for about 15 minutes or until the smell of ether is gone. Split the tube contents in two and add 2.25 times the sample volume of 100% ethanol to each tube, mix the tube contents by inverting the tube several times, and leave at room temperature (22 °C) for 30 minutes.
- Microfuge the tube at 16,000 g for 45 minutes and discard the supernatant.
- Add 200 µl of 70% ethanol to gently wash the pellet and then microfuge at 16,000 g for 15 minutes.
- Discard the supernatant and dry the pellet at 56 °C for 30–45 minutes with the tube lid open.
- Resuspend the DNA in 30 µl distilled water that is RNAase and DNAase free.
- Store the sample tube at 4 °C until tested (few days) or at –20 °C for long-term storage.

Test Method

Lower reaction mixtures for the DVEV PCR and the lambda control are prepared in advance in a biosafety cabinet using the kit manufacturer's recommended methods for a hot start PCR. The lower reaction mixture is dispensed into tubes, sealed with wax at 80 °C, as recommended by the manufacturer, and stored at 4 °C for 1–2 months.

PCR primers for DVEV DNA-directed DNA polymerase gene.

Primer 1 sequence: 5'-GAA-GGC-GGG-TAT-GTA-ATG-TA-3' (forward).

Primer 2 sequence: 5'-CAA-GGC-TCT-ATT-CGG-TAA-TG-3' (reverse).

- The upper reaction mixture is prepared according to the kit manufacturer's recommendations the day of the test, and distributed to each sample tube including DVEV and lambda control tubes.
- Add 10 µl of DNA suspension from the stored sample tubes to the PCR lower reaction tubes with corresponding labels.
- Place known DVEV DNA diluted to 1 pg/10 µl into one control tube and 10 µl of distilled water into the no DNA control tube. Add 10 µl of lambda DNA supplied in the kit and 10 µl of water to the corresponding lambda control tubes.
- Place all the tubes in a thermal cycler that is programmed as follows:

One cycle: Hold 94 °C for 2 minutes.

Hold 37 °C for 1 minute.

Hold 72 °C for 3 minutes.

35 cycles: Hold 94 °C for 1 minute.

Hold 55 °C for 1 minute.

Hold 72 °C for 2 minutes.

One cycle: Hold 72 °C for 7 minutes.

Hold 4 °C until stored.

PCR tubes are stored at 4 °C until the samples are examined for amplification products.

Electrophoretic Analysis of PCR Products

- A fresh 1 × TAE buffer (40 mM Tris acetate, 1 mM EDTA, pH 8.3) is prepared from a 10× stock for agarose preparation and for use in the electrophoresis chamber.
- A 1% agarose solution is prepared in TAE buffer, heated to dissolve the agarose, and, when cool, poured into a gel former with a comb.
- The solidified gel is placed into the electrophoresis chamber and TAE running buffer is added.
- PCR test samples, including the DVEV and lambda controls, are mixed 1/10 with 1 µl of loading buffer (0.25% [w/v] bromophenol blue, 0.25% [w/v] xylene cyanol, 0.01 M Tris/HCl, pH 8.0, and 50% [v/v] glycerol) and 10 µl of each is added to individual wells of the gel. The 100 bp molecular size markers are added to each side of the gel.
- Run the gel for 1 hour at 120 volts and then stain in a 1% ethidium bromide solution for 20 minutes (alternative, safer stains should preferably be used to visualise PCR products). De-stain the gel for 45 minutes in deionised water and view the gel on a UV-illuminated light box. Photograph the gel to record results.

Interpretation of the Results

A 500 bp amplification band in the lambda control sample indicates the PCR ran successfully. A 446 bp band in the DVEV known DNA control indicates the DVEV primers are working. A 446 bp band in the unknown test sample indicates DVE viral DNA was present. No amplification products will be present in the DVEV or lambda no DNA controls. If bands appear in these negative control products, cross-contamination occurred during the test set-up and the test must be repeated.

Real-time PCR

The following is an example protocol for a real-time PCR for detection of DVEV; other protocols exist.

PCR primers targeting a 124-bp fragment of the DVE DNA polymerase gene.

Primer 1 sequence: 5'-CTC-TAC-GCA-GCT-TTT-GAC-GAT-TT-3' (forward).

Primer 2 sequence: 5'-AGA-AAC-ATA-CTG-TGA-GAG-TGA-CGA-3' (reverse).

The labelled probe (5'-CCT-CCT-CCT-CGC-TGA-GTG-GCA-TCC-3') is complementary to a 24 bp region between the upstream and downstream primer pair labelled with 6-carboxyfluorescein at the 5' end and 6-carboxy-tetramethyl-rhodamine at the 3' end.

- The DVEV DNA extraction can be performed using a suitable tissue/cell DNA extraction kit or the extraction procedure. The PCR can be performed using a suitable real-time PCR kit and PCR detection system and software.
- The amplification is performed in a total of 25 µl PCR mixture containing 1 µl of DNA solution, 13PCR buffer, 10 mM MgCl2 , 0.4 mM dNTP, 0.2 µM concentration of each primer, 0.24 µM fluorogenic probe, and 1.25 U of Taq polymerase.
- The PCR conditions consisted of: a) one cycle of 5 minutes at 95 C b) 40 two-step cycles of 5 seconds at 94 °C and 20 seconds at 65 °C.
- If quantification is desired, the number of target copies in the reaction can be deduced from the threshold cycle (CT) values corresponding to the fractional cycle number at which the released fluorescence exceeds 15 times the standard deviation of the mean baseline emission.

Strain Variation

Although strains of DVEV differ considerably in virulence, there is little reported evidence of serological variation.

Serological Tests

Serological tests have little value in the diagnosis of acute DVE infections, but assays based on serum neutralisation in embryonated eggs and cell cultures have been used to monitor antibodies following exposure to DVE in wild waterfowl. The humoral response to natural infection with DVEV is often low and antibodies may be short-lived; it is assumed that cell-mediated immunity also plays a role in the infection. However, detection of neutralising antibodies to DVEV in serum is possible. Virus neutralisation (VN) assays using a constant-serum/varying-virus method may be performed in chicken or duck embryos by using embryo-adapted virus, or in cell cultures. For laboratories lacking duck embryos, serological diagnosis is possible by virus neutralisation, using a chicken embryo fibroblast adapted DVEV strain and primary chicken embryo fibroblasts (CEF). Neutralisation indices (NI) between 0 and 1.5 were detected in domestic and wild waterfowl that had not been exposed to DVEV; a NI of 1.75 or greater was considered to be evidence of prior exposure to DVEV. Alternatively, sera may be screened using a constant-virus/varying-serum method. In the author's laboratory a microtitre neutralisation assay using primary MDEF or DEF is used. Serial twofold dilutions of each serum sample (heatinactivated at 56 °C) are prepared in 50 µl of serum-free MEM in microtitre plates. Approximately $10^{2.0}$ $TCID_{50}$ (50% tissue culture infective dose) of DVEV in 50 µl of MEM is added to each well and the mixtures are allowed to react at 37 °C for 1 hour. A suspension of primary MDEF or DEF in MEM supplemented with 2 mM L-glutamine, 0.17% sodium bicarbonate and 10% FCS, are adjusted to contain 3×10^5 cells per ml. Cells are next added to the plates at 100 µl per well and the plates are then incubated for up to 96 hours at 37 °C in a humidified 5% CO_2 atmosphere. Following incubation, cells are observed daily by light microscopy and finally fixed with 10% formolbuffered saline and stained with 1% crystal violet. The plates are read macroscopically. The titre for virus neutralis-ing activity is expressed as the reciprocal of the highest dilution of serum at which a monolayer grew, i.e. there is no evidence of CPE and therefore complete virus neu-tralisation has occurred. A titre of less than 3 $\log_2$ is usually considered to be negative. A VN titre of 8 or greater is considered to be significant and is evidence of exposure to DVEV. VN antibody may also be detected using cell cultures by mixing sera at a single dilution, e.g. 1/10, with 100–200 $TCID_{50}$ virus and then testing inoculated cell cultures for nonneutralised virus by immunofluorescence. Although this method is not quantitative, it can be useful for screening large numbers of sera. These latter methods, using constant-virus/varying-serum, are much more economical on sera than the NI methods.

Immunochromatographic (ICS) strip tests have been developed for DVE antibody detection and may have promise for use in the field. The ICS strip test is based on membrane chromatography and uses recombinant UL51 protein as the capture antigen. This test is reported to have a sensitivity comparable to ELISA and much higher than VN tests. A dot-ELISA and passive haemagglutination tests have been reported for the detection of DVE antibodies but the sensitivity and specificity of these tests are moderate. Several indirect ELISAs have been described for the serological detection of DVE. Indirect ELISA using the entire DVEV virion as coated antigen has been described for DVE antibody detection and is commercially available. Several indirect ELISAs that use recombinant DVE proteins that act as coating antigen have also been developed. Wu described an indirect ELISA using a recombinant UL55 protein of DVE expressed in E. coli (UL55-ELISA). Compared with a commercial indirect ELISA based on whole DVE virions and VN tests, the UL55-ELISA was found to be intermediate in sensitivity and specificity. Another indirect ELISA using thymidine kinase fusion protein expression in E. coli as coating antigen was reported to detect post-vaccine DVE antibodies 5 days earlier compared with conventional assays. While ELISAs based on recombinant DVE proteins are reported to be rapid, simple and more economical alternatives for DVE serological detection, further testing of their stability is needed before widespread use.

Requirements for Vaccines

Characteristics of the Seed

Biological Characteristics

DVE vaccine can be prepared from a strain of the virus that has been attenuated by serial passage in embryonated chicken eggs. In the USA the vaccine strain seed was originally imported from Holland and has been serially passaged 41–46 times.

The seed virus should be prepared in 8- to 11-day-old specific pathogen free (SPF) embryonated chicken eggs by inoculating on to the CAM followed by incubation at 37 °C. The seed may be stored at −70 °C or lower in the form of a homogenate of the embryo CAM in buffered saline

Quality Criteria (Sterility, Purity, Freedom, from Extraneous Agents)

The seed virus should be shown to be free from extraneous viruses pathogenic to ducks, chickens and turkeys. It should also be free from bacterial, fungal and mycoplasmal contaminants.

The identity of the virus should be confirmed by a VN test conducted with specific antiserum using the constant-serum/varying-virus method. This test should be performed in embryonated chicken eggs. The antiserum should reduce the virus titre by at least 101.75 ELD50 (50% embryo lethal dose).

Methods of Manufacture

Procedure

The vaccine is produced in 8–11-day-old SPF embryonated chicken eggs inoculated on to the CAM and incubated at 37 °C. Most embryo deaths occur between 48 and 96 hours after inoculation. The embryos, their CAMs and chorioallantoic fluids are harvested, pooled and homogenised in buffered saline and clarified by low-speed centrifugation (1800 g). The preparation is diluted as appropriate, and a stabiliser is incorporated. It is then dispensed into vials and preferably frozen rapidly to −70 °C or lower.

Requirements for Substrates and Media

All reagents should be sterile and eggs obtained from a specific pathogen-free source.

In Process Controls

Any embryo dying within the first 24 hours of inoculation should be discarded as non-specific deaths.

Final Product Batch Tests

- Sterility and purity:

Tests for sterility and freedom from contamination of biological materials intended for veterinary use.

- Safety:

Safety tests in target animals are not required by many regulatory authorities for the release of each batch. Where required, standard procedures are generally conducted using fewer animals than are used in the safety tests required for the relevant regulatory approval. A group of 1-day-old ducklings susceptible to DVE should be inoculated subcutaneously or intramuscularly with the vaccine at 10 times the recommended dose, and observed for 7– 14 days for any signs of adverse reactions.

- Batch potency:

The virus titre of the vaccine should be determined in 9- to 11-day-old embryonated chicken eggs inoculated on to the CAM and incubated at 37 °C. The vaccine should contain a minimum of 103.0 ELD_{50} per dose at time of use. The immunogenicity of the vaccine can be assessed in DVE-susceptible ducks or ducklings by inoculating the recommended vaccine dose intramuscularly and challenging intramuscularly 21 days later with virulent DVEV. The vaccinated birds should survive challenge while unvaccinated control birds should die. This test should be carried out on the master seed but need not be done routinely on each vaccine batch produced. For release of subsequent batches, the titre of the virus should be a sufficient indication of vaccine potency.

Requirements for Authorisation

Safety Requirements

A group of 1-day-old ducklings susceptible to DVE should be inoculated subcutaneously or intramuscularly with the vaccine at 10 times the recommended dose, and observed for 7– 14 days for any signs of adverse reactions.

- Target and non-target animal safety:

The vaccine is intended solely for use to protect ducklings and ducks against DVEV.

- Reversion-to-virulence for attenuated/live vaccines:

There are no reports of reversion to virulence by the DVE vaccine.

- Environmental consideration:

None.

Efficacy Requirements

- For animal production:

Immunity in vaccinated ducks should last throughout a breeding season. Annual revaccination is recommended.

- For control and eradication:

The vaccine virus is not thought to spread by contact from vaccinated to unvaccinated ducks, as the unvaccinated birds remain susceptible to infection.

Stability

When stored at –70 °C or lower the vaccine is stable for at least 1 year. Potency testing should be repeated after this time on an aliquot of vaccine to determine whether virus titre has been maintained. Once thawed the vaccine should not be refrozen, it should be maintained at 4 °C in a refrigerator but for no longer than 1 week. Lyophilised vaccine should be stored at 4–8 °C and used before the stated expiry date.

Vaccines based on Biotechnology

Vaccines Available and their Advantages

Research has been published on the development and efficacy of recombinant DVE vaccines in SPF ducks. In 2011, Liu reported the use of a DVEV-vectored live bivalent vaccine in which the haemagglutinin gene of H5N1 avian influenza virus was inserted between the unique short (US) 7 and US8 genes of the DVEV genome. This bivalent vaccine was reported to be efficacious against both DVEV and H5N1 avian influenza viral infection in SPF ducks under experimental conditions. DVE vaccines based on biotechnology show promise under experimental conditions but are currently not commercially available for widespread use.

Oncorhynchus Masou Virus Disease

Oncorhynchus masou virus disease (OMVD) is an oncogenic and skin ulcerative condition coupled with hepatitis among salmonid fish in Japan, and probably in the coastal rivers of eastern Asia that harbour Pacific salmon.

Disease Information

Aetiological Agent and Agent Strains

The aetiological agent is Oncorhynchus masou virus (OMV) in the family Herpesviridae, although it has also been given the name nerka virus Towada Lake, Akita and Amori Prefecture (NeVTA), yamame tumour virus (YTV), coho salmon tumour virus (CSTV, COTV), O. kisutch virus (OKV), coho salmon herpesvirus (CHV), rainbow trout kidney virus (RKV), and rainbow trout herpesvirus (RHV).

Survival Outside the Host

A significant reduction in the infectious titre of OMV was observed within 3 and 7 days in environmental water at 15 °C and 10 °C, respectively. However, the infectivity remained for 7–14 days below 5 °C indicating evidence for the presence of bacterial strains in the water with anti-viral activity.

Stability of the Agent (Describe Effective Inactivation Methods)

Pathogen-free water sources are often essential in aquaculture. Water that comes from rivers or lakes, commonly used in hatcheries, contains fish pathogens. Such open water supplies should not be used without treatment to kill fish pathogens. Fish viruses are divided into two groups based on sensitivity to UV. OMV belongs to a sensitive group and is inactivated by treatment with 104 μW second cm^{-2} ultraviolet dose. At 15 °C for 20 minutes, minimum concentrations showing 100% plaque reduction of OMV by iodophor, sodium hypochlorite solution, benzalkonium chloride solution, saponated cresol solution, formaldehyde solution and potassium permanganate solution were 40, 50, 100, 100, 3500 and 16 mg $litre^{-1}$, respectively.

OMV is heat-, ether-, and acid (pH 3)-labile and does not haemagglutinate human O-cells. It is completely inactivated by ultraviolet (UV) irradiation with 3.0×10^3 μW second cm^{-2}. In the presence of 50 μg ml^{-1} of the pyrimidine analogue, 5-lododeoxyuridine (IUdR), replication is inhibited. Replication of OMV is also inhibited by anti-herpesvirus agents such as phosphonoacetate (PA), acyclovir (ACV), (E)-5-(2-bromovinyl)-2'-deoxyuridine (BVdU), and 1-B-D-arabino- furanosylcytosine (Ara-C), which is caused by the inhibition of DNA polymerase induced by OMV.

Life Cycle

Following the septicaemia phase of OMV infection, an immune response takes place that results in the synthesis of neutralising antibodies to OMV. A carrier state frequently occurs that leads to virus shedding via the sexual products at the time of spawning.

Host Factors

Susceptible Host Species (Common and Latin Names)

Fish species that are susceptible to OMV include: Kokanee (sockeye) salmon (Oncorhynchus nerka), masu salmon (O. masou), chum salmon (O. keta), coho salmon (O. kisutch) and rainbow trout (O. mykiss).

Susceptible Stages of the Host

The age of the fish is critical and 1-month-old alevins are the most susceptible target for virus infection. The main environmental factor favouring OMV infection is low water temperature, below 15 °C.

Species or Subpopulation Predilection (Probability of Detection)

Salmonids are the only fish species susceptible to OMV infection, the order of the fish species from the most to the least susceptible is kokanee salmon, chum salmon, masou salmon, coho salmon and rainbow trout.

Target Organs and Infected Tissue

Clinically, the initial infection by OMV appears as a systemic and frequently lethal infection that is associated with oedema and haemorrhages. Virus multiplication in endothelial cells of blood capillaries, haematopoietic tissue and hepatocytes underlies the clinical signs. Four months after this first clinical condition, a varying number of surviving fish exhibit epithelioma occurring mainly around the mouth (upper and lower jaw) and, to a lesser extent, on the caudal fin, operculum and body surface. This neoplasia may persist for up to 1 year post-infection. In the case of coho salmon, 1-year-old infected fish in particular show ulcers on the skin, white spots on the liver and neoplastic tissues around the mouth parts or body surface. In rainbow trout, commercial sized fish were infected and the diseased fish exhibit almost no external signs, although some fish manifest ulcerative lesions on the skin. Internally, intestinal haemorrhage and white spots on the liver are observed.

Persistent Infection with Lifelong Carriers

Under natural conditions, survivors of OMVD are persistently infected with virus and they shed the virus and fish retain the virus until maturation.

Vectors

Water is the major abiotic vector. However, animate vectors, e.g. other fish species, parasitic invertebrates and piscivorous birds and mammals may also be involved in transmission.

Known or Suspected Wild Aquatic Animal Carriers

Masu salmon that were caught at the river mouth had neoplasia around the mouth and OMV was isolated from tumours. Recently rainbow trout that are living in the river and that may have escaped from farms were infected with OMV and died.

Disease Pattern

Transmission Mechanisms

The reservoirs of OMV are clinically infected fish and covert carriers among groups of cultured, feral or wild fish. Infectious virus is shed via faeces, urine, sexual products and probably skin mucus, while the kidney, spleen, liver and tumours are the sites where virus is the most abundant during the course of overt infection. The transmission of OMV is horizontal and possibly 'egg-surface associated'. Horizontal transmission may be direct or vectorial, water being the major abiotic factor. Disinfection of the eggs just after fertilisation and eyed stage is effective in preventing OMV infection. OMV disease was not reported in alevins originating from disinfected eggs that had been incubated and hatched in virus-free water.

Prevalence

OMV was isolated from masu salmon at all the investigated sites with the exception of one hatchery. Based on our epizootiological study, the roots of OMV was assumed to be along the Japan Sea coast of Hokkaido and presumed original host species was masu salmon. In the 1960s, eggs of masu salmon were collected from the rivers of Japan Sea coast of Hokkaido, and transported to Honshu Island, main land of Japan. With the unrestricted fish movement, the virus spread to several places in Honshu where the first cancer disease of masu salmon was observed. Subsequently, coho salmon and rainbow trout were cultured in the same water systems where masu salmon was cultured. Coho salmon might be infected with OMV at fry stage in fresh water because tumour tissues were found around the mouth of pen cultured coho salmon, the hatchery from where coho salmon was transplanted to pen had a history of OMVD.

Geographical Distribution

Following the first reports of OMVD in Northern Japan, the geographical range of the disease has become extensive inside Japan. There are no reports of disease occurrence outside Japan.

Mortality and Morbidity

Susceptibility of several salmonid fry to OMV has been studied experimentally by immersion in water containing 100 $TCID_{50}$ (median tissue culture infective dose) ml^{-1} OMV at 10 °C for 1 hour. Comparing the five different salmonid fry, at the age of

1 month, kokanee salmon exhibited the greatest sensitivity with 100% mortality. Masu and chum salmon also exhibited high sensitivity at 87% and 83% mortality, respectively. Coho salmon and rainbow trout were shown to be less sensitive to OMV infection at 39% and 29% mortality, respectively. Thus the host range of OMV is wide in salmonid species. Eight age groups of chum salmon, (0, 1, 2, 3, 4, 5, 6, and 7 months old), were immersed under the same conditions. The cumulative mortality of just hatching chum salmon, observed in ensuing 4 months, was 35%, but between 1-month and 5-month-old fry, it was more than 80%. At 6 and 7 months, the fry's susceptibility was reduced and only 7%, and 2% fish had succumbed to the disease. There were no deaths among 8-month-old fingerlings. On the other hand, 1-month-old masu salmon fry was most sensitive and the cumulative mortality reached 87%. In 3- to 5-month-old fry, cumulative mortality decreased from 65% to 24%.

Since 1988, herpesvirus had been isolated from the liver, kidney, and developing neoplasm in pond and pen-cultured coho salmon. Affected fish showed the following disease signs, ulcers on their skin, white spots on their liver and neoplastic tissues around their mouth part or body surface. Coho salmon culture is economically damaged by this disease. All of these viruses were neutralised by anti-OMV or NeVTA rabbit serum, and the oncogenicity was confirmed by experimental infection. Isolated virus showed strong pathogenicity to coho salmon. Massive mortality has occurred among 1-year-old rainbow trout in pond cultures since 1992 in Hokkaido. The diseased fish exhibited almost no clinical signs. Some fish did manifest ulcerative lesions on their skin. Internally, intestinal haemorrhage and white spots on the liver were observed. Epizootics occurred in cultured rainbow trout weighing 12 g to 1.5 kg at 18 fish farms from February 2000 to January 2001 in Nagano Prefecture, Japan. High infectivity titres about 10^8 $TCID_{50}$ g^{-1} were demonstrated in the main internal organs and multiple necrotic foci were observed in the liver. The virus was identified as OMV using serological tests and polymerase chain reaction (PCR). In more than 80% cases, the outbreaks were linked with introductions of live fish.

Environmental Factors

General sanitation measures are standard practice in hatcheries. Special care must be taken to avoid the movement of equipment from one tank to another and all should be disinfected after use. Methods to sanitise a hatching unit should be carefully developed with respect to chemical toxicity for fish, effects of water temperature and their repeated use. It should be remembered that workers themselves might serve as efficient vectors for pathogens and proper disinfection of hands and boots are required to prevent dissemination of viruses. Although it may be difficult to sanitise hatching and rearing units during use, raceways and ponds should be disinfected with chlorine before and after use.

Control and Prevention

OMV is sensitive to ultraviolet irradiation, ozone or iodophor treatment. Since 1983, it has been strongly recommended as a control strategy that inspection of the ovarian fluid from mature fish and the disinfection of collected eggs in all hatcheries in Hokkaido with iodine at the early eyed stage be done. Currently OMV is no longer detected in most of the hatcheries in this area. Nowadays, all eggs and facilities had been disinfected by iodophor just after fertilization and again at the early eyed stage. As a result, OMV cannot be isolated in Hokkaido and Tohoku area, and could avoid the outbreak of OMVD of masu salmon and coho salmon except rainbow trout.

Vaccination

Vaccination of mature rainbow trout with formalin-inactivated OMV could reduce the positive ratio of OMV in ovarian fluid. Also vaccination using formalin-inactivated OMV is very effective to protect the OMV infection at the fry stage.

Chemotherapy

The therapeutic efficacy of ACV was evaluated using OMV and chum salmon fry. The fish were experimentally infected with OMV, and were treated with ACV either orally or by immersion. Daily immersion of fish into ACV solution (25 μg ml^{-1}, 30 minutes per day, 15 times) reduced mortality of the infected fish. Oral administration of the ACV (25 μg per fish per day, 60 times) did not affect survival of the chum salmon. On the contrary, the group administered IUdR by the oral route showed a higher survival than the ACV-administered group. This suggested that an effective level of ACV was not maintained in fish given the drug by the oral route. Daily immersion of infected fish into ACV solution (25 μg ml^{-1}, 30 minutes per day, 60 times) considerably suppressed the development of tumours induced by OMV.

Immunostimulation

There is currently no published information on the use of immunostimulants to control OMVD in salmonids. However, it is known to be an area of research interest.

Resistance Breeding

There is currently no published information on the use of resistance breeding to control OMVD in salmonids.

Restocking with Resistant Species

Hybrids represent a potential control method to prevent serious losses from OMVD. Studies on a population of triploid hybrid salmonid (tetramer rainbow trout × brown trout) found them to be resistant to OMVD.

Blocking Agents

Not applicable.

Disinfection of Eggs and Larvae

Disinfection of eggs can be achieved by iodophor treatment. OMV has been shown to be inactivated by iodophor at 50 mg litre^{-1} for 15 minutes at 15 °C or 25 mg litre^{-1} for 20 minutes at 15 °C.

General Husbandry Practices

Biosecurity measures should include ensuring that new introductions of fish are from disease-free sources and installation of a quarantine system where new fish can be held with sentinel fish at permissive temperatures for OMVD. The fish are then quarantined for a minimum of 4 weeks to 2 months before transfer to the main site and mixing with naïve fish. Hygiene measures on site should be similar to those recommended for IHN and include disinfection of eggs, regular disinfection of ponds, chemical disinfection of farm equipment, careful handling of fish to avoid stress and safe disposal of dead fish.

Sampling

Selection of Individual Specimens

Clinically Affected Fish

Whole alevin (body length ≤ 4 cm), viscera including kidney (4 cm ≤ body length ≤ 6 cm) or, for larger size fish, skin ulcerative lesions or neoplastic tissues, and kidney, spleen, liver and encephalon.

Apparently Healthy Fish

Kidney, spleen and encephalon (any size fish) and ovarian fluid from broodfish at spawning time.

Preservation of Samples for Submission

Whole fish should be sent to the laboratory alive or killed and packed separately in sealed aseptic containers. However, it is highly preferable and recommended to collect organ samples from the fish immediately after they have been selected at the fish production site. Whole fish or selected organ samples should be sent to the laboratory in refrigerated containers (+0 °C to 5 °C) with ice. The freezing of collected fish or dissected organs should be avoided.

Pooling of Samples

When testing clinically affected fish by cell culture method or PCR-based methods, pooling of samples should be avoided or restricted to a maximum of five fish per pool. For health surveillance testing by cell culture methods, samples should be tested in a maximum of five fish per pool.

Best Organs or Tissues

Clinically Affected Fish

Whole alevin (body length ≤ 4 cm), viscera including liver or kidney (4 cm ≤ body length ≤ 6 cm) or, for larger size fish, skin ulcerative lesions or neoplastic tissues, and liver or kidney.

Apparently Healthy Fish

Liver, kidney, spleen and encephalon (any size fish) and ovarian fluid from brood fish at spawning time.

Samples/Tissues that are not Suitable

Fish carcasses showing very advanced signs of tissue decomposition may not be suitable for testing by any methods.

Diagnostic Methods

Field Diagnostic Methods

Clinical Signs

Clinically, the initial infection by OMV appears as a systemic and frequently lethal infection that is associated with oedema and haemorrhages. Virus multiplication in endothelial cells of blood capillaries, haematopoietic tissue and hepatocytes underlies the clinical signs. Four months after this first clinical condition, a varying number of surviving fish exhibit epithelioma occurring mainly around the mouth (upper and lower jaw) and, to a lesser extent, on the caudal fin, operculum and body surface. This neoplasia may persist for up to 1 year post-infection. In the case of coho salmon, 1-year-old infected fish in particular show ulcers on the skin, white spots on the liver and neoplastic tissues around the mouth parts or body surface. In rainbow trout, the diseased fish exhibit almost no external signs, although some fish manifest ulcerative lesions on the skin. Internally, intestinal haemorrhage and white spots on the liver are observed.

Behavioural Changes

Fish become lethargic, gather at the water outlet or sides of a pond. Some fish may experience loss of equilibrium and disorientation.

Clinical Methods

Gross Pathology

Gross signs in infected fish are inappetence and exophthalmia, and petechiae on the body surface, especially beneath the lower jaw. Agonal or abnormal swimming behaviour has not been observed. Internally, the liver shows white spots lesions, and in advanced cases the whole liver becomes pearly white. In some cases the spleen is found to be swollen. The digestive tract is devoid of food.

Microscopic Pathology

The kidney of OMV-infected 1- and 3-month old masu salmon, 1-month-old coho salmon and 2-month-old chum salmon is the principal target organ for the virus as judged by the severity of histopathological changes found in infected 1-month-old masu salmon. Necrosis of epithelial cells and kidney were observed in the early moribund specimens while partial necrosis of the liver, spleen and pancreas was seen in later moribund specimens from this group. Necrosis of the kidney haematopoietic tissue was observed in infected 3-month-old masu salmon. While, the kidney was considered to be the early target organ for OMV, it gradually became resistant to OMV infection. For this reason, it was considered that the principal target organ moved from the kidney to the liver and marked histopathological changes were observed in the later stages. Foci of necrosis in the liver tended to become more severe with a longer incubation period. Hepatocytes showing margination of chromatin were present. Cell degeneration in the spleen, pancreas, cardiac muscle and brain was also observed. Histopathological changes observed in coho salmon and chum salmon were the same as those of masu salmon. In the case of rainbow trout, high infectivity titres were demonstrated in the main internal organs and multiple necrotic foci were observed in the liver. The definite change was necrosis of OMV-infected cells, which were observed in the spleen, haematopoietic tissues in the kidney, liver, intestine, heart, gill filaments, epidermis and lateral musculature. In particular, the intestines showed severe necrosis and haemorrhage in the epithelium and underlying tissues, which is the new description of rainbow trout OMVD.

Wet Mounts

OMV has been identified in touch imprints of kidney by indirect fluorescence antibody test (IFAT).

Electron Microscopy/Cytopathology

Viral particles have been detected by transmission electron microscopy (TEM) examination of liver tissues from clinically infected chum salmon, masu salmon, coho salmon and rainbow trout. Electron microscopy of infected cells reveals that the intra nuclear hexagonal capsids have a diameter of 115 nm. Abundance of budding, enveloped virions, 200 × 240 nm in diameter, are also observed on the surface and inside cytoplasmic vesicles. The calculated number of capsomeres of negatively stained virions is 162. These features confirm that OMV is a herpesvirus.

Agent Detection and Identification Methods

The agent's infectivity remains unchanged for at least 2 weeks at 0 °C to 5 °C, but at –20 °C, 99.9% of the infectivity is lost within 17 days. Virus isolation should be carried out using fish, transported on ice to the laboratory. For filtration of the OMV, a 0.40 μm nucleopore filter (polycarbonate) is recommended because cellulose acetate membrane filter traps virus particles. For the purpose of a virological survey of mature salmonid, ovary fluid is collected by the method described by Yoshimizu, 1985, with the addition of the same volume of antibiotic and reacted at 5 °C, overnight. In the case of the tumour tissue, tissue is cut and disinfected with iodophor, then washed with Hank's BSS and transported with antibiotic solution to the laboratory. Tumour tissues must be prepared for the primary culture or co-culture with RTG-2 cells. After the one subculture of primary culture cells, the virus inspection of the culture medium should be carried out. Usually RTG-2 cells are harvested and inoculated, suitable incubation temperature is 15 °C. In the laboratory, rabbit serum or monoclonal antibody against OMV was used for a fluorescent antibody test, and also DNA probe was used for detection of virus genome. PCR using a F10 primer and R05 primer amplified a 439 base-pair segment of DNA from OMV strains isolated from masu salmon, coho salmon and rainbow trout, and liver, kidney, brain and nervous tissues. Agarose gel profile of amplified DNA was able to distinguish OMV and H. salmonis.

Direct Detection Methods

OMV has been identified in touch imprints of kidney by IFAT. The most commonly used method for detection of OMV directly in fish tissues is using PCR method specific for OMV.

Microscopic Methods

Wet Mounts

Virus antigens have been detected in infected tissues by IFAT. In the case of coho salmon, pond cultured fish are transplanted to net pens in the sea. Kidney tissues were pressured hard to adapt to the marine environment. During this period, OMV replicate and OMV antigen appeared in kidney tissues. Indirect fluorescent antibody method is useful and effective to detect the OMV infected fish.

Fixed Sections

The method detailed above is also suitable for detection of OMV antigen in paraffin wax tissue sections fixed in 10% neutral buffered formalin (NBF). A common treatment is incubation of the sections with 0.1% trypsin in PBS at 37 °C for 30 minutes. The sections are then washed in cold PBS.

Agent Isolation and Identification

Cell Culture/Artificial Media

Cell line to be used: RTG-2 or CHSE-214.

Inoculation of Cell Monolayers

- Make an additional tenfold dilution of the 1/10 organ homogenate supernatants and transfer an appropriate volume of each of the two dilutions on to 24-hour-old cell monolayers. Inoculate at least 2 cm^2 of drained cell monolayer with 100 µl of each dilution.
- Allow to adsorb for 1 hour at 15 °C and, without withdrawing the inoculate, add cell culture medium buffered at pH 7.4 and supplemented with 2% fetal bovine serum (FBS) (1 ml/well for 24-well cell culture plates), and incubate at 15 °C.

Monitoring Incubation

- Follow the course of infection in positive controls and other inoculated cell cultures by microscopic examination at ×4 or 10 magnification for 14 days. The use of a phase-contrast microscope is recommended.
- Maintain the pH of the cell culture medium at between 7.2 and 7.4 during incubation. This can be achieved by the addition of sterile bicarbonate buffer (for tightly closed cell culture flasks) or Tris buffer solution (for cell culture plates) to the inoculated cell culture medium or, even better, by using HEPES-buffered medium (HEPES = N-2-hydroxyethyl-piperazine-N-2-ethanesulfonic acid).
- If a cytopathic effect (CPE) appears in those cell cultures inoculated with the dilutions of the tested homogenate supernatants, identification procedures have to be undertaken immediately. If a fish health surveillance/control programme is being implemented, provisions may have to be taken to suspend the approved health status of the production unit or the zone from which the virus-positive sample originated. The suspension of approved status will be maintained until it is demonstrated that the virus in question is not OMV.
- If no CPE develops in the inoculated cultures (despite normal progression of CPE in the virus controls), the inoculated cultures should be subcultured for a further 7 days. Should the virus controls fail to develop CPE, the process should be repeated with fresh susceptible cells and new batches of samples.

Subcultivation Procedures

- Collect aliquots of cell culture medium from all monolayers inoculated with dilutions of each supernatant of organ homogenates.

- If required, repeat the neutralisation test to infectious pancreatic necrosis virus (IPNV) and infectious haematopoietic necrosis virus (IHNV) as previously described, with a dilution of the above supernatant (1/1 to 1/100).
- Inoculate cell monolayers as described above.
- Incubate and monitor as described above.
- If no CPE occurs, the test may be declared negative.

Isolation of OMV from Cultures of Neoplastic Cells

- Collect neoplastic tissues, disinfect with iodophor, 50 parts per million for 20 minutes, and wash three times with Hanks' balanced salt solution.
- The tissues are left overnight in 0.25% trypsin in phosphate buffered saline (PBS) at 5 °C. Then, 3.5 × 105 neoplastic cells/ml are seeded in a tissue culture flask and incubated with culture medium containing 20% fetal bovine serum (FBS).
- Harvest the primary neoplastic cell culture and co-cultivate with RTG-2 or CHSE-214 cells.
- Incubate and monitor as described above.

Antibody-based Antigen Detection Methods

Neutralisation Test

- Collect the culture medium of the cell monolayer exhibiting CPE and centrifuge at 2000 g for 15 minutes at 4 °C to remove cell debris.
- Dilute the virus-containing medium from 102 to 104 ml^{-1}.
- Mix aliquots (for example 200 µl) of each virus dilution with equal volumes of an antibody solution specific for OMV, and similarly treat aliquots of each virus dilution with cell culture medium. (The neutralising antibody [NAb] solution must have a 50% plaque reduction titre of at least 2000.)
- In parallel, other neutralisation tests must be performed against:
    - A homologous virus strain (positive neutralisation test).
    - A heterologous virus strain (negative neutralisation test). v) If required, a similar neutralisation test may be performed using antibodies to IPNV, to ensure that no IPNV contaminant has escaped the first anti-IPNV test.
- Incubate all the mixtures at 15 °C for 1 hour.

- Transfer aliquots of each of the above mixtures on to cell monolayers (inoculate two cell cultures per dilution) and allow adsorption to occur for 0.5~1 hour at 15 °C; 24- or 12-well cell culture plates are suitable for this purpose, using a 50 µl inoculum.
- When adsorption is completed, add cell culture medium supplemented with 2% FCS and buffered at pH 7.4~7.6 into each well and incubate at 10~15 °C.
- Check the cell cultures for the onset of CPE and read the results as soon as it occurs in non-neutralised controls (cell monolayers being protected in positive neutralisation controls). Results are recorded either after a simple microscopic examination (phase-contrast preferable) or after discarding the cell culture medium and staining the cell monolayers with a solution of 1% crystal violet in 20% ethanol.
- The tested virus is identified as OMV when CPE is prevented or noticeably delayed in the cell cultures that had received the virus suspension treated with the OMV-specific antibody, whereas CPE is evident in all other cell cultures.
- In the absence of any neutralisation by NAb to OMV, it is mandatory to conduct an IFAT with the suspect sample.

Molecular Techniques

PCR

- Extract nucleic acid from cells infected with OMV strain OO-7812 and H. salmonis using the InstGene Matrix (Biorad2).
- Pellet the virus-infected tissues or infected cultured cells by centrifugation at 19,000 g for 15 minutes.
- Wash the pellets twice with 1 ml PBS and mix with 200 µl of chelating resin (Sigma).
- Incubate the mixture at 56 °C for 20 minutes in a water bath, vortex it, and then place it in a boiling water bath for 8 minutes.
- Vortex the samples and centrifuge at 8200 g (10,000 rpm) for 90 seconds.
- Subject the supernatant to PCR.
- The forward primer (F10) is 5'-GTA-CCG-AAA-CTC-CCG-AGT-C-3', and the reverse primer (R5) is 5'-AAC-TTG-AAC-TAC-TCC-GGG-G-3'.
- Incubate the specimens, primer sets and reaction mixtures for 30 cycles in an automatic thermal cycler (GeneAmp PCR 9700, Applied Biosystems), with each

cycle consisting of denaturation at 94 °C for 30 seconds, annealing at 56 °C for 30 seconds and extension at 72 °C for 30 seconds.

- Analyse the amplified product for size and purity by electrophoresis (100 V for 30 minutes) in 2% agarose gel and stain with ethidium bromide.
- A PCR using these primer sets amplified a 439 base-pair segment of DNA from OMV strains isolated from masu salmon, coho salmon and rainbow trout, and liver, kidney, brain and nervous tissues, and an 800 base-pair segment of DNA from SalHV-1. SalHV-1 and SalHV-2 could be distinguished by agarose gel profile of this amplified DNA.

Serological Methods

Indirect Fluorescent Antibody Test

- Prepare monolayers of cells in 2 cm² wells of cell culture plastic plates or on cover-slips in order to reach around 80% confluency, which is usually achieved within 4 hours of incubation at 22 °C (seed six cell monolayers per virus isolate to be identified, plus two for positive and two for negative controls). The FBS content of the cell culture medium can be reduced to 2–4%. If numerous virus isolates have to be identified, the use of Terasaki plates is strongly recommended.
- When the cell monolayers are ready for infection, i.e. on the same day or on the day after seeding, inoculate the virus suspensions to be identified by making tenfold dilution steps directly in the cell culture wells or flasks.
- Dilute the control virus suspension of OMV in a similar way, in order to obtain a virus titre of about 5000-10,000 plaque-forming units (PFU) ml^{-1} in the cell culture medium.
- Incubate at 15 °C for 48 hours. v) Remove the cell culture medium, rinse once with 0.01 M PBS, pH 7.2, then three times briefly with cold acetone (stored at −20 °C) for cover-slips or a mixture of acetone 30%/ethanol 70%, also at −20 °C, for plastic wells.
- Let the fixative act for 15 minutes. A volume of 0.5 ml is adequate for 2 cm² of cell monolayer.
- Allow the cell monolayers to air-dry for at least 30 minutes and process immediately or freeze at −20 °C.
- Prepare a solution of purified antibody or serum to OMV in 0.01 M PBS, pH 7.2, containing 0.05% Tween 80 (PBST), at the appropriate dilution (which has been established previously or is given by the reagent supplier).

- Rehydrate the cell monolayers by four rinsing steps with the PBST solution, and remove this buffer completely after the last rinsing.
- Treat the cell monolayers with the antibody solution for 1 hour at 37 °C in a humid chamber and do not allow evaporation to occur. The volume of solution to be used is 0.25 ml per 2 cm^2 well.
- Rinse four times with PBST as above.
- Treat the cell monolayers for 1 hour at 37 °C with a solution of fluorescein isothiocyanate (FITC)-conjugated antibody to the immunoglobulin used in the first layer and prepared according to the instructions of the supplier. These FITC antibodies are most often rabbit or goat antibodies.
- Rinse four times with PBST.
- Examine the treated cell monolayers on plastic plates immediately, or mount the cover-slips using glycerol saline at pH 8.5 prior to microscopic observation.
- Examine under incident UV light using a microscope with ×10 eye pieces and ×20–40 objective lens having numerical aperture > 0.65 and > 1.3, respectively. Positive and negative controls must be found to give the expected results prior to any other observation.

Enzyme-linked Immunosorbent Assay

- Coat the wells of microplates designed for enzyme-linked immunosorbent assays (ELISAs) with appropriate dilutions of monoclonal antibody or purified immunoglobulins (Ig) specific for OMV, in 0.01 M PBS, pH 7.2 (200 µl/well).
- Incubate overnight at 4 °C.
- Rinse four times with 0.01 M PBS containing 0.05% Tween 20 (PBST).
- Block with skim milk (5% in PBST) or other blocking solution for 1 hour at 37 °C (200 µl/well).
- Rinse four times with PBST.
- Add 2% Triton X-100 to the virus suspension to be identified.
- Dispense 100 µl/well of a two- or four-step dilution of the virus to be identified and of OMV control virus, and allow to react with the coated antibody to OMV for 1 hour at 20 °C.
- Rinse four times with PBST.

- Add to the wells, biotinylated polyclonal antibody to OMV.
- Incubate for 1 hour at 37 °C.
- Rinse four times with PBST.
- Add streptavidin-conjugated horseradish peroxidase to those wells that have received the biotin-conjugated antibody, and incubate for 1 hour at 20 °C.
- Rinse four times with PBST.
- Add the substrate and chromogen. Stop the course of the test when positive controls react, and monitor the results.

Rating of Tests against Purpose of Use

As an example, the methods currently available for targeted surveillance and diagnosis of OMVD are listed in table. The designations used in the table indicate: a = the method is the recommended method for reasons of availability, utility, and diagnostic specificity and sensitivity; b = the method is a standard method with good diagnostic sensitivity and specificity; c = the method has application in some situations, but cost, accuracy, or other factors severely limits its application; and d = the method is presently not recommended for this purpose. These are somewhat subjective as suitability involves issues of reliability, sensitivity, specificity and utility. Although not all of the tests listed as category A or B have undergone formal standardisation and validation, their routine nature and the fact that they have been used widely without dubious results, makes them acceptable.

Table: Methods for targeted surveillance and diagnosis.

Method	Targeted surveillance				Presumptive diagnosis	Confirmatory diagnosis
Gross signs	d	b	b	b	b	c
Cell culture	b	b	b	b	b	a
Direct LM	d	d	d	d	d	d

Table: Methods for targeted surveillance and diagnosis.

Method	Targeted surveillance				Presumptive diagnosis	Confirmatory diagnosis
	Larvae	PLs	Juveniles	Adults		
Hispathology	b	b	b	b	b	a
Transmission EM	c	c	c	c	c	c
Antibody-based assays	b	b	b	b	b	a
DNA Probes in situ	c	c	c	c	c	c
PCR	b	b	b	b	a	a
Sequence	d	d	d	D	d	d

PLs = postlarvae; LM = light microscopy; EM = electron microscopy; PCR = polymerase chain reaction.

Tests Recommended for Targeted Surveillance to Declare Freedom from Oncorhynchus Masou Virus Disease

Information on the distribution and incidence of OMV is important for the prevention of transmission to the progeny from mature salmonids. Therefore, study of the occurrence of OMV among mature salmonid fish is important. Sixty fish were sampled and specimens were collected individually. Ovarian fluid specimens were collected according to the method of Yoshimizu. A sterilised automatic pipette tip was inserted into the urogenital opening of the mature fish. One milliliter of ovarian fluid taken from the fish was treated by the antibiotic treatment method. Antibiotic treated specimens were transported to the laboratory with ice.

Corroborative Diagnostic Criteria

Suspect Case

OMV shall be suspected if at least one of the following criteria is met:

- The presence of typical clinical signs of OMVD in a population of susceptible fish.
- Presentation of typical histopathology in liver tissue sections consistent with OMVD.
- A single positive result from one of the diagnostic assays such as IFAT on liver or kidney tissue imprints or PCR.
- Transfer of live fish from a site where presence of OMV has been confirmed, or is suspected, because of the presence of clinical disease, to sites without suspicion of OMV.
- Antibodies to OMV have been detected.

Confirmed Case

The following criteria should be met for confirmation of OMV:

- Mortality, clinical signs and pathological changes consistent with OMV disease and detection of OMV by one or more of the following methods:
    - Isolation and identification of OMV in cell culture from at least one sample from any fish on the site as.
    - Detection of OMV by PCR by the methods.

    - Detection of OMV in tissue preparations by means of specific antibodies against OMV (e.g. IFAT on tissue imprints as.

- In the absence of mortality or clinical signs by one or more of the following methods:

    - Detection and confirmation of OMV by PCR by the methods.

    - Positive results from two separate and different diagnostic assays described above.

References

- Herpesviridae: wikiwand.com, Retrieved 16 April, 2019
- Bovine-Herpesvirus-1: microbewiki.kenyon.edu, Retrieved 23 July, 2019
- Equine-herpesvirus, farms-fishing-forestry-agriculture-livestock-animal-welfare-pests-diseases-disorders: business.qld.gov.au, Retrieved 14 March, 2019
- Suid-herpesvirus, agricultural-and-biological-sciences -1: sciencedirect.com, Retrieved 21 July, 2019
- Detection and identification of Oncorhynchus masou virus (OMV) disease by polymerase chain reaction (PCR). Bull. Fish. Sci. Hokkaido Univ., 52, 111–116.

4

Understanding Adenoviridae Viruses

A type of virus, of size ranging between 90-100nm, having no outer envelope with double stranded DNA genome is referred to as adenoviridae virus. Canine Adenovirus, Hydropericardium Syndrome, Egg Drop Syndrome Virus, etc. are some of the viruses that fall in its domain. The topics elaborated in this chapter will help in gaining a better perspective about adenoviridae viruses.

Adenoviridae

Adenovirus is any virus belonging to the family Adenoviridae. This group of viruses was discovered in the 1950s and includes 6 genera and 47 species (formerly referred to as serotypes) that cause sore throat and fever in humans, hepatitis in dogs, and several diseases in fowl, mice, cattle, pigs, and monkeys. The virus particle lacks an outer envelope; is spheroidal, about 80 nm (1 nm = 10^{-9} metre) across; is covered with 252 regularly arranged protein subunits called capsomeres; and has a core of double-stranded deoxyribonucleic acid (DNA) wrapped in a protective coat of protein. Adenoviruses develop within the nuclei of infected cells, where they are often observed packed in an apparently crystalline arrangement.

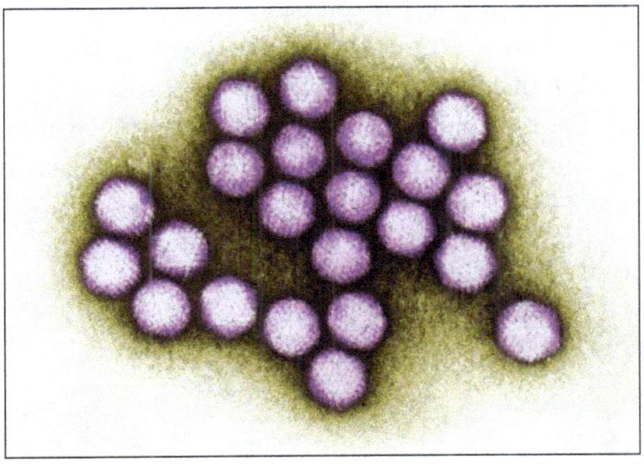

Adenovirus, colourized transmission electron micrograph.

In humans, adenoviruses cause acute mucous membrane infections of the upper respiratory tract, the eyes, and frequently the regional lymph nodes, bearing considerable resemblance to the common cold. Adenoviruses can also cause epidemic keratoconjunctivitis (EKC) and are considered to be responsible for an outbreak of respiratory disease among military recruits in 1997. Like the cold viruses, adenoviruses are often found in latent infections in clinically healthy persons. Of the 47 different adenovirus species, only a few commonly cause illness in humans; it is thus possible to prepare a vaccine against these viruses. Vaccines include a first-generation inactivated vaccine against several adenovirus types and a non-attenuated oral vaccine against adenovirus types 4 and 7. In contrast, there are more than 100 cold viruses, all of which are commonly found as disease agents; this great number makes the development of a vaccine for the common cold virtually impossible.

Adenoviruses are highly environmentally resistant viruses that often cause illness of the gastrointestinal tract and liver. This virus, which is also referred to as Atadenovirus, "wasting disease," or "star-gazing disease," has been known to spread to the kidneys, brain, and bone marrow in reptiles.

Within reptiles, each adenovirus tends to infect only one species or group (i.e., cross-infection and zoonosis are not likely). Bearded dragons have a very high chance of having this disease. Transmission is not completely understood, but definitely includes fecal-oral routes, and transmission from mother to offspring. It presents the most danger to very young, very old, or otherwise immune compromised patients. Unfortunately, the virus is never really cleared from the animal, so positive animals can shed the virus for life, potentially giving it to other members of the species. Healthy carrier adults may show no clinical signs whatsoever.

Clinical signs of adenovirus in affected reptiles include:

- Failure to thrive.
- Anorexia & weight loss.
- Lethargy.
- Green feces or urates.
- Hind leg weakness.
- Abnormal postures.

- Neurologic deficiencies.
- Parasite or bacterial infections that will not go away.
- Sudden death.

While many of these symptoms are not unique to this virus, these clinical signs warrant consideration of this virus as a potential cause of ongoing disease.

Testing involves a swab of the oral surfaces, cloacal surfaces and feces being sent to an outside lab for PCR testing (which looks for virus DNA). Results usually return in 7-10 days. At this point in time, there is no need to retest a negative patient unless it comes into contact with a "status unknown" or "positive" bearded dragon. Positive animals can lead a very productive life, but should be kept apart from negative animals.

There is no vaccine for any reptile adenovirus. Treatment in patients showing signs of illness consists of supportive care, including: fluid therapy, assist feeding, liver support medications, anti-inflammatory medication, and (if available) anti-viral medications such as oseltamvir (Tami-Flu). Increasing vegetables in the diet may help as well. Again, it presents the most danger to very young, old, or otherwise immune-compromised patients. If you are considering getting a bearded dragon as a pet, try to get one that is over 3-6 months old.

Canine Adenovirus

Infectious canine hepatitis (ICH) is a worldwide, contagious disease of dogs with signs that vary from a slight fever and congestion of the mucous membranes to severe depression, marked leukopenia, and coagulation disorders. It also is seen in foxes, wolves, coyotes, bears, lynx, and some pinnipeds; other carnivores may become infected without developing clinical illness. In recent years, the disease has become uncommon in areas where routine immunization is done, but periodic outbreaks, which may reflect maintenance of the disease in wild and feral hosts, reinforce the need for continued vaccination.

ICH is caused by a nonenveloped DNA virus, canine adenovirus 1 (CAV-1), which is antigenically related only to CAV-2 (one of the causes of infectious canine tracheobronchitis. CAV-1 is resistant to lipid solvents (such as ether), as well as to acid and formalin. It survives outside the host for weeks or months, but a 1%–3% solution of sodium hypochlorite (household bleach) is an effective disinfectant.

Ingestion of urine, feces, or saliva of infected dogs is the main route of infection. Recovered dogs shed virus in their urine for ≥6 mo. Initial infection occurs in the tonsillar crypts and Peyer patches, followed by viremia and disseminated infection. Vascular endothelial cells are the primary target, with hepatic and renal parenchyma, spleen, and lungs becoming infected as well. Chronic kidney lesions and corneal clouding ("blue eye") result from immune-complex reactions after recovery from acute or subclinical disease.

Clinical Findings

Signs vary from a slight fever to death. The mortality rate ranges from 10%–30% and is typically highest in very young dogs. Concurrent parvoviral or distemper infection worsens the prognosis. The incubation period is 4–9 days. The first sign is a fever of >104°F (40 °C), which lasts 1–6 days and is usually biphasic. If the fever is of short duration, leukopenia may be the only other sign, but if it persists for >1 day, acute illness develops.

Signs are apathy, anorexia, thirst, conjunctivitis, serous discharge from the eyes and nose, and occasionally abdominal pain and vomiting. Intense hyperemia or petechiae of the oral mucosa, as well as enlarged tonsils, may be seen. Tachycardia out of proportion to the fever may occur. There may be subcutaneous edema of the head, neck, and trunk. Despite hepatic involvement, there is a notable absence of icterus in most acute clinical cases.

Clotting time is directly correlated with the severity of illness and is the result of disseminated intravascular coagulation induced by vascular endothelial compromise, coupled with failure of the liver to rapidly replace consumed clotting factors. It may be difficult to control hemorrhage, which is manifest by bleeding around deciduous teeth and by spontaneous hematomas. CNS involvement is unusual and is typically the result of vascular injury. Severely infected dogs may develop convulsions from forebrain damage. Paresis may result from brain stem hemorrhages, and ataxia and central blindness have also been described. Foxes more consistently have CNS signs and intermittent convulsions during the course of illness, and paralysis may involve one or more limbs or the entire body. Respiratory signs usually are not seen in dogs with ICH; however, CAV-1 has been recovered from dogs with signs of infectious tracheobronchitis despite high serologic titers against parenteral disease.

Clinicopathologic findings reflect the coagulopathy (prolonged prothrombin time, thrombocytopenia, and increased fibrin degradation products). Severely affected dogs show acute hepatocellular injury (increased ALT and AST). Proteinuria is common. Leukopenia typically persists throughout the febrile period. The degree of leukopenia varies and seems to be correlated with the severity of illness.

On recovery, dogs eat well but regain weight slowly. Hepatic transaminase activities peak around day 14 of infection and then decline slowly. In ~25% of recovered dogs, bilateral corneal opacity develops 7–10 days after acute signs disappear and usually resolves spontaneously. In mild cases, transient corneal opacity may be the only sign of disease.

It has long been thought that chronic hepatitis may develop in dogs that have low levels of passive antibody when exposed, although a recent PCR-based study did not confirm this theory.

Lesions

Endothelial damage results in "paint-brush" hemorrhages on the gastric serosa, lymph nodes, thymus, pancreas, and subcutaneous tissues. Hepatic cell necrosis produces a variegated color change in the liver, which may be normal in size or swollen. Histologically, there is centrilobular necrosis, with neutrophilic and monocytic infiltration, and hepatocellular intranuclear inclusions. The gallbladder wall is typically edematous and thickened; edema of the thymus may be found. Grayish white foci may be seen in the kidney cortex.

Diagnosis

Usually, the abrupt onset of illness and bleeding suggest ICH, although clinical evidence is not always sufficient to differentiate ICH from distemper. Definitive antemortem diagnosis is not required before institution of supportive care but can be pursued with commercially available ELISA, serologic, and PCR testing. PCR or restriction fragment length polymorphism is required to definitively distinguish CAV-1 from CAV-2, if clinically necessary. Postmortem gross changes in the liver and gallbladder are more conclusive, and diagnosis is confirmed by virus isolation, immunofluorescence, characteristic intranuclear inclusion bodies in the liver, or PCR or fluorescence in situ hybridization studies of infected tissue.

Treatment

Treatment is symptomatic and supportive. The goals of therapy are to limit secondary bacterial invasion, support fluid balance, and control hemorrhagic tendencies. Broad-spectrum antibiotics and intravenously administered balanced electrolyte solutions with 5% dextrose supplementation are indicated. Plasma or whole blood transfusions may be necessary in severely ill dogs.

Although the transient corneal opacity (which may be seen during the course of ICH or associated with vaccination with attenuated CAV-1 vaccines) usually requires no treatment, atropine ophthalmic ointment may alleviate the painful ciliary spasm sometimes associated with it. Dogs with corneal clouding should be protected against bright light. Systemic corticosteroids are contraindicated for treatment of corneal opacity associated with ICH.

Prevention

Modified-live virus (MLV) injectable vaccines are available and are often combined with other vaccines. Vaccination against ICH is recommended at the time of canine distemper vaccinations. Maternal antibody from immune bitches interferes with active immunization in puppies until they are 9–12 wk old. Attenuated CAV-1 vaccines have produced transient unilateral or bilateral opacities of the cornea, and the virus may be shed in urine. For these reasons, CAV-2 attenuated live virus strains, which provide cross-protection against CAV-1, are preferentially used because they have very little tendency to produce corneal opacities or uveitis, and the virus is not shed in urine. Historically, annual revaccination against ICH was standard, and vaccines are labeled for annual use. Increasing evidence suggests that immunity induced by MLV CAV-1 injectable vaccines lasts ≥3 yr, although this remains an extra-label use of commercially available vaccines.

Hydropericardium Syndrome

From a veterinary viewpoint, the avian adenoviruses can be divided into three groups, i.e. groups I, II and III.

Group I, or conventional adenoviruses, share a common group antigen, distinct from the mammalian adenovirus group antigen. These viruses grow readily in avian cell cultures and have been isolated from chickens, turkeys, geese, ducks, quail, pigeons, ostriches and other avian species. Fowl adenoviruses can be divided into at least twelve serotypes. A major problem in classification has been the presence of prime strains and strains of broad antigenicity. Five groups (A-E) have also been distinguished on the basis of restriction endonuclease analysis using two enzymes. The fowl adenoviruses not only infect chickens, but also turkeys and many other species. Turkeys, geese and ducks are affected by adenoviruses that do not grow or only grow poorly in chicken cell cultures and require a homologous cell type. At least three serotypes have been isolated from turkeys, and these grow in turkey but not chicken cells. A study of the relationship between isolates found in the United States of America (USA) and Northern Ireland, and between these turkey isolates and other avian strains, remains to be undertaken.

Three serotypes have been isolated from geese and one from Muscovy ducks (Cairina moschata).

Group II adenoviruses include the viruses of turkey haemorrhagic enteritis (THE), marble spleen disease (MSD) and group II splenomegaly of chickens. These viruses share a common antigen which is distinct from the group antigen of mammalian and group I avian adenoviruses.

Group III viruses, the egg drop syndrome (EDS) viruses, are widely distributed in waterfowl but can easily infect chickens, resulting in the production of abnormal eggshells.

Until recently, two genera have been recognised within the family Adenoviridae, namely: Mastadenovirus (mammalian strains including human strains) and Aviadenovirus.

A third genus has recently been proposed, the genus Atadenovirus. Egg drop syndrome virus would be included in the Atadenovirus genus, together with bovine adenovirus 5, 6, 7 and 8, and ovine adenovirus isolate 287. The position of the avian group II (THE/ MSD) viruses in this classification is unclear.

Aetiology

The adenovirus virion is a non-enveloped icosahedral particle of 70 nm-90 nm in diameter. The particle has 252 capsomeres arranged in twelve triangular faces with six capsomeres along each edge. The nucleic acid is linear, double-stranded deoxyribonucleic acid. The virions have a density in caesium chloride of between 1.32 g/ml and 1.37 g/ ml. Adenoviruses replicate in the nucleus, producing basophilic inclusions.

All adenoviruses are resistant to lipid solvents, sodium deoxycholate, trypsin, 2% phenol and 50% alcohol. They are resistant to exposure at pH 3 to pH 9, but are inactivated by 1:1,000 formalin. The avian adenoviruses appear to be more resistant to thermal inactivation than mammalian adenoviruses. Some strains survive 60 °C and even 70 °C for 30 min, and an F l isolate was reported to survive 18 h at 56 °C. At present, information on the effect of divalent cations is conflicting. Most workers accept that divalent cations destabilise adenoviruses, but some studies found no effect. Within the group 1 adenoviruses, only some strains of F l agglutinate rat erythrocytes.

Group I (Conventional) Adenoviruses

Epidemiology and Pathogenesis

Adenoviruses are ubiquitous in chickens, as demonstrated by serological surveys and virological studies, and have been isolated from both sick and healthy birds. Adenoviruses have also been isolated from turkeys, geese, ducks, pigeons, budgerigars and a mallard duck (Anas platyrhynchos). Evidence of adenovirus infection has been recorded in gulls, psittacines, owls and hawks. Infection by adenoviruses is likely to occur in all species of birds.

Transmission

Vertical transmission is a very important route. Chicks hatching from infected eggs may excrete virus in faeces from the rime of hatching, but more typically chicks do not excrete virus until two to four weeks of age. Presumably reactivation of latent virus does not occur until maternal antibody declines. In a broiler flock where chicks originate from different parent flocks, a massive interchange of strains occurs, and concurrent infections of one bird with two or even three serotypes is not unusual. Spread of virus in this way results in peak virus excretion in a flock between four and six weeks. In one study of a layer replacement flock, virus excretion was at a maximum between five and nine weeks, but 70% of birds were still excreting after fourteen weeks. In another study, virus excretion again remained at a high level until fourteen weeks, and eight different serotypes were isolated from seven farms. Birds can re-excrete virus throughout life. Following a period of excretion, the virus appears to become latent, presumably due to the development of local immunity. When the local immunity is lost, after eight to twelve weeks, the virus is unmasked and excretion occurs. Humoral antibody does not appear to play a role in preventing excretion, as adult birds have been found to excrete virus despite high levels of neutralising antibody to the same serotype. Humoral antibody appears to offer little or no protection against infection with a different serotype. Adenoviruses are frequently isolated from hens during the period of peak egg production. This upsurge in virus activity ensures maximum transmission of virus to the next generation, through the egg.

Horizontal transmission is also important. The virus is excreted in high titres in the faeces. In addition, virus grows in the nasal and tracheal mucosa, conjunctiva and kidneys, and therefore virus could be present in other secretions or excretions. Virus could also be present in semen, which could be important where artificial insemination is used. Excretion of virus in the faeces follows a different pattern in juveniles and adults. In the juvenile, higher titres of virus are excreted for longer periods than in the adult. Lateral spread appears to occur principally by direct contact between birds or indirect contact by people, crates, egg trays and trolleys. Airborne spread probably only occurs over very short distances. True aerosol spread between farms is highly unlikely, but virus in contaminated poultry litter from a depopulated house could present a risk. In broiler houses, infection appears to spread very rapidly, but this is probably due to reactivation of latent virus in many birds throughout the house. When introduction of virus is minimal, as in a specific-pathogen-free (SPF) flock, spread can be very slow.

Disease

A wide range of virulence has been reported within the adenoviruses and the viruses are ubiquitous. Many infections are subclinical, in some cases because birds still have some maternal immunity when infected, but in many cases because the viruses have low virulence. The lack of virulence of some strains is illustrated by the fact that many SPF flocks become infected, even during lay, without any signs being observed. However, because latent adenovirus infections often become apparent at approximately two to three weeks of age, and again around peak egg production (i.e. during periods when disease or production problems are rife), adenoviruses have been associated with a range of conditions such as respiratory disease, diarrhoea, reduced egg production, detrimental effects on feed conversion and arthritis. In most of these conditions, the role of the adenovirus, if any, is that of a helper or secondary pathogen, rather than a primary pathogen. Thus, a study in Denmark was unable to detect any effect of adenoviruses on broiler flock performance. However, adenovirus is an important pathogen in some outbreaks of disease.

Inclusion Body Hepatitis

Inclusion body hepatitis (IBH) is usually seen in meat-producing birds between three and seven weeks of age, but has also been recorded in birds as young as seven days, and as old as twenty weeks. Classically, IBH is associated with sudden onset mortality which peaks within three to four days and ceases by days five to six, although in some outbreaks, deaths have continued for up to three weeks. Morbidity is low. Affected birds crouch, have ruffled feathers and die or recover within 48 h. Mortality usually ranges between 5% and 10%, but can reach 30%. Within an integrated breeding organisation, disease episodes in broiler flocks have been associated with certain breeder flocks.

The liver is the primary organ affected. Some reports suggest that the target organ is the haemopoietic system, but the aplastic anaemia described was probably due to simultaneous infection with chicken anaemia virus. The liver is pale, swollen and friable, and petechial or ecchymotic haemorrhages may be present. Haemorrhages may also be present in the musculature. Numerous eosinophilic intranuclear inclusions, and infrequently basophilic inclusions, are found in the hepatocytes. For many years, the role of adenoviruses in IBH has been unclear. Many serotypes have been associated with outbreaks of IBH. Adenoviruses are observed in the basophilic inclusions, but the eosinophilic inclusions are composed of fibrillar granular material. Experimental reproduction of IBH using adenoviruses has been inconclusive. Most workers have had no success, but some experimental infections have produced liver lesions and death following parenteral inoculation. However, the hepatocyte nuclei contained basophilic inclusions, rather than the eosinophilic inclusions typical of natural outbreaks.

Recent outbreaks of IBH have been described in Australia in birds under three weeks of age. Mortality was up to 30% and basophilic nuclear inclusions predominated in the hepatocytes. Reproduction of the condition was possible using serotypes 6, 7 and 8 isolated from field cases, administered by natural routes. All isolates were genetically closely-related, possessing a group E genotype. The field isolates were further divided into hypervirulent and mildly pathogenic isolates, using nine endonucleases. Recombination studies indicated that the fibre was responsible for the differences in virulence between isolates.

The serotypes isolated from severe outbreaks of IBH in New Zealand were principally F8 and also F1 and F12. In addition to the liver lesions where eosinophilic inclusions predominated, atrophy of the bursa and thymus was reported, together with aplastic bone marrow. These isolates all belonged to genotype E, but were distinct from the genotype found in Australia.

Necrotising pancreatitis and intranuclear inclusions have been observed in natural cases of IBH, and pancreatitis has occurred in experimentally infected chickens. Gizzard erosions and ulceration were present, but no intranuclear inclusion bodies were detected in the gizzard epithelial cells in outbreaks of IBH. Focal necrotising pancreatitis and gizzard erosions with typical adenovirus inclusions containing virus particles in necrotic pancreatic acinar cells and gizzard epithelial cells have also been seen in the absence of IBH. The latter birds were also infected with chicken anaemia virus. Other workers have also noted gizzard erosions, necrotising pancreatitis and mild proventriculitis with wet unformed faeces, in birds orally infected with adenovirus Infection with infectious bursal disease virus (IBDV) has been suggested as a major predisposing factor in the development of IBH. However, in New Zealand, and in the early cases in Northern Ireland, IBDV was absent. Furthermore, spontaneous IBH has been reported in SPF birds free of IBDV.

Adenoviral IBH has been recorded in pigeons, kestrels and a merlin (Falco columbarius), and in day-old turkeys from which turkey adenovirus serotype 2 was recovered. Pancreatitis was also found in some of the pigeons.

Hydropericardium Syndrome

In 1987, a new syndrome named hydropericardium syndrome (HPS) or Angara disease was recognised in Pakistan. The disease has subsequently been recognised in India, Kuwait, Iraq, Mexico, Central and South America, Japan and Russia. The disease in Central and South America has been diagnosed as IBH. Hydropericardium syndrome differs from IBH only in that the mortality rate and the incidence of hydropericardium are much higher.

The disease principally affects meat-producing birds between three and six weeks of age, with mortality from 20 % to 80%. Hydropericardium syndrome also occurs in breeding and laying flocks, with lower mortality rates. The disease is characterised by the accumulation of clear fluid (up to 10 ml) in the pericardium. Pulmonary oedema, an enlarged liver and pale enlarged kidneys are usually present. In addition, multifocal coagulative necrosis of the liver is observed, with mononuclear cell infiltration and intranuclear basophilic inclusions in the hepatocytes. The serological response to Newcastle disease vaccination is impaired.

The disease is considered to be the result of infection with adenovirus type 4 or 8 although some workers consider that other factors may be involved.

An HPS-like disease has been reported in pigeons, and broilers injected with liver from affected pigeons developed.

Disease in Turkeys

Adenoviruses have been isolated from clinical outbreaks of respiratory disease, diarrhoea and depressed egg production and more recently, IBH in day-old turkeys. Attempts to reproduce the diseases have generally been unsuccessful.

Disease in Waterfowl

Three serotypes isolated from geese failed to reproduce disease in experimentally infected goslings. In a disease outbreak with high mortality associated with hepatitis, adenovirus-like particles were observed in the live.

In Canada, an isolated parent flock produced two hatches in which mortality in four- to eleven-day-old goslings reached 12% due to respiratory tract disease. A diptheritic stenosing tracheitis with occasional bronchitis and pneumonia, in which tracheal epithelial cells contained numerous adenovirus particles, was reported in 10% of seven- to twenty-one-day-old Muscovy ducks.

Disease in Guinea-fowl

Pancreatitis and focal pancreatic necrosis with large basophilic and smaller eosinophilic inclusions have been associated with adenoviral infection of guinea-fowl. Pancreatitis and respiratory lesions have been induced by intranasal inoculation of adenovirus into day-old guinea-fowl. A haemorrhagic disease of guinea-fowl in which adenoviral inclusions were present in the spleen has been reported and reproduced experimentally.

Disease in Ostriches

Adenoviruses have been associated with illness, diarrhoea, pancreatitis, death and poor hatchability in ostriches. An isolate from an ostrich produced pancreatitis in guinea-fowl. In a study where three-day-old ostrich chicks were inoculated with an ostrich-derived adenovirus, all inoculates died.

Quail Bronchitis

Quail bronchitis is an acute, highly contagious disease of young bobwhite quail (Colinus virginianus). Disease is most severe in one- to three-week-old birds, with morbidity approaching 100% and mortality up to 50%. Antibody has been detected in older birds and in wild quail. Disease has also been seen in Japanese quail (Coturnix coturnix japonica).

Quail bronchitis is caused by a type 1 fowl adenovirus which is indistinguishable from chicken isolates. No information is available regarding whether the F l strain behaves in quail as it does in chickens, where latency and vertical transmission occur. Chickens and turkeys may be experimentally infected with isolates from quail, but develop only very mild symptoms of disease.

Gross lesions in quail bronchitis include evidence of ocular and nasal discharge, mucoid tracheitis and airsacculitis. Occasionally, haemorrhagic exudate is present in the trachea. Histologically, a necrotising tracheitis, proliferative and necrotising bronchitis and pneumonia are observed. Basophilic intranuclear inclusions are common in tracheal epithelial cells. Multifocal necrotising hepatitis, splenitis and bursal lymphoid necrosis leading to atrophy are also seen.

Other Diseases in Quai

Two cases of adenoviral inclusion body ventriculitis have been diagnosed in bobwhite quail. In coturnix quail, gastrointestinal disease with inclusions in the digestive tract, particularly in the caeca, has been reported recently.

Diagnosis

A detailed methodology has been described in the literature for group I adenoviruses and for quail bronchitis.

Virus Isolation

The preferred sample is faeces or colon with faeces. If a particular organ has obvious lesions, for example, the liver in IBH, or the trachea in quail bronchitis, this should also be included. Virus is frequently present in bursa of Fabricius, nasal mucosa, pharynx, trachea, lung and kidney. A 10% suspension of the specimen is made in cell culture media or bacteriological broth. In both cases, antimicrobial agents such as 1,000 international units (IU) of penicillin/ml and 1,000 µg streptomycin/ml should be added. The suspensions can be stored at 4 °C or -20 °C or below until required. Isolation is usually undertaken in cell cultures. For chickens, chick embryo liver or chick kidney cells are best. Chick embryo fibroblasts are insensitive and chick embryo liver cell cultures must show a predominance of epithelial cells. These cells are also suitable for preliminary isolation attempts from other species. However, some adenoviruses that affect turkeys, and probably other avian species, only grow in the cells of homologous species. Therefore, where possible, the homologous cell type should be used, for example, turkey kidney when investigating turkeys. One difficulty is the lack of SPF eggs for most species other than chickens. Because of the widespread distribution of adenoviruses and the presence of virus in eggs, an SPF source is virtually essential. If unavailable, SPF chicken eggs may be the only choice. Following inoculation, the cell cultures should be observed for fourteen days before being discarded. This usually involves one blind passage. Uninoculated cells should be treated in the same manner, to check for the presence of latent virus. Both rolled cultures and flasks are equally sensitive. Frequently, more than one adenovirus serotype, or more than one virus is isolated, for example adenovirus and reovirus. To acquire pure cultures, the use of plaque purification or the limiting dilution techniques often associated with the use of • neutralising antisera is necessary.

If adenovirus is present, round cells which detach from the glass are observed. As a routine practice, all isolates should be checked for the presence of haemagglutinins, to exclude Orthomyxoviridae and Paramyxoviridae. Adenoviruses of group I and II do not agglutinate fowl erythrocytes. The most rapid method of confirming the presence of adenovirus is indirect immunofluorescence. If available, direct examination of disrupted cell preparations with the electron microscope is also a rapid method of recognition, as the virus morphology is typical. However, if the serotype is to be established, the isolate must be typed against the standard antisera.

Embryonated eggs, inoculated by the allantoic route are not sensitive, except in the case of virus types 1 and 5. Laboratory isolates have been successfully propagated in eggs following inoculation of the yolk sac.

Modern biochemical methods can be employed, but are of limited value. Polymerase chain reaction (PCR) techniques may be inappropriate because latency makes it impossible to determine if a positive result is due to the disease currently being investigated or an earlier infection. However, genotyping may be a valuable tool to distinguish between pathogenic and non-pathogenic strains.

Serological Detection

The double immunodiffusion (gel precipitation) test has been widely used. However, the low cost in materials and labour is probably the only advantage of this test. The main disadvantages are lack of sensitivity and detection of group antigen. The test has been used widely to monitor SPF flocks for freedom from adenovirus infection where only group antigen detection is required. However, in many cases, the test has remained negative when birds in SPF flocks have become infected. This has been confirmed by experimental studies which have demonstrated that birds undergoing a primary infection as a result of natural exposure may not respond with precipitin antibodies. The apparent sensitivity of the test in the field is a result of the birds being infected with two or more strains. The sensitivity of the test can be increased by using a pool of antigen prepared from three different serotypes.

The test of choice to monitor SPF flocks is the enzyme-linked immunosorbent assay (ELISA). Little benefit is derived from using a test to detect group antibody in commercial birds, given the widespread extent of infection.

The serum neutralisation test is used to detect type-specific antibody. This is time consuming and expensive, even using the microtitre technique, because a minimum of twelve serotypes must be used when testing chicken sera.

Public Health Importance

Group I adenoviruses do not naturally infect mammals and therefore no public health implications exist.

Prevention and Control

The widespread distribution of group I adenoviruses throughout the world means that eradication would not be possible. Furthermore, some strains may be able to move between domestic and wild birds. Until recently, development of vaccines has not been a priority because of the absence of important diseases caused by adenoviruses. Since the recent outbreaks of IBH and HPS, development of vaccines has been attempted with varying success. A formalin inactivated liver suspension with liquid paraffin adjuvant is reported to be highly effective against HPS. Some other inactivated vaccines have also given good results.

No trade implications exist for infections with conventional adenoviruses. Obviously, movement of birds or eggs from flocks infected with the highly virulent viruses associated with HPS or the recent outbreaks of IBH to uninfected areas would not be wise. However, at present, testing for these conditions is not possible. Thus, type 8 viruses belonging to restriction enzyme group E have been associated with new variant IBH, but similar viruses have also been isolated from normal, healthy birds. The best option is certification that the birds, or in the case of eggs, the parents, have not demonstrated signs of HPS or new variant IBH.

Group II Adenoviruses

Group Il has three known members, namely: turkey haemorrhagic enteritis virus (THEV), marble spleen disease virus (MSDV) and avian adenovirus group II splenomegaly virus (AASV) of chickens. These viruses share a common antigen, which is distinct from that shared by the group I or conventional avian adenoviruses, and from mammalian adenoviruses.

Convalescent THEV serum protects pheasants against MSD. A single serological type of group 11 viruses appears to exist and isolates are classified only as to the source (e.g. THEV or MSDV). Isolates can be distinguished from one another by restriction endonuclease analysis and monoclonal antibody affinity.

lnfectivity resists heating for 1 h at 65 °C, but is destroyed after 1 h at 70 °C. The viruses demonstrate a wide range of virulence, ranging from highly virulent to non-virulent.

Culture in conventional cell cultures such as turkey kidney or chick embryo liver is not possible. Growth occurs in a turkey lymphoblastoid B cell line derived from a Marek's disease induced tumour, the MDTC-RPI9 cell line. Virus has also been grown in turkey peripheral blood leukocytes.

Disease

Turkey Haemorrhagic Enteritis

Turkey haemorrhagic enteritis virus is distributed widely throughout the world. Antibody studies demonstrate that a high proportion of adult domestic turkeys have been infected, although a study of wild turkeys reported no evidence of infection. Guinea-fowl and psittacines may be naturally infected. Other gallinaceous birds such as peafowl, bobwhite quail and chukars can be infected. Lesions develop in the latter, but deaths have not been reported. A serological survey of forty-two species of wild birds indicated no evidence of a reservoir outside the Galliformes.

Turkey haemorrhagic enteritis usually occurs .in turkeys between six and eleven weeks, although a case has been described in 2.5-week-old poults. Turkeys under thirteen days old appear to be resistant to infection in the absence of maternal immunity, presumably because target cells have not adequately matured. No upper age limit exists for infection Transmission is faecal-oral. Virus is present in faeces for several weeks and further bursts of excretion may occur when local antibody wanes. The virus is very rsistant and can easily be carried from farm to farm by humans. Infection is also liable to recur in successive flocks in the same house, unless cleansing and disinfection is meticulous. No evidence of egg transmission has been found.

The virus replicates initially in the lymphoid cells of the intestinal tract and bursa of Fabricius. Virus can be detected one day post infection (dpi), peaks at 4 dpi-7 dpi and remains detectable up to 15 dpi in the intestinal tract. Virus is recoverable from the bursa between 2 dpi and 7 dpi. Virus is present in plasma from 2 dpi, and virus replicates are detected in the blood leukocytes from 3 dpi to 18 dpi. The spleen is the major site of viral replication. Antigen is detectable in the spleen from 2 dpi, reaches a peak at 6 dpi and is no longer detectable at 18 dpi. Reports as to the amount of antigen in the intestine are conflicting, and the intestinal pathology may be immune-mediated. Apoptosis occurs in approximately half of the immunoglobulin M+ cells at 3 dpi but not in cluster of differentiation 4+ (CD4+) and CD8+ T lymphocytes, and occurrence of apoptosis is not restricted to infected cells. The rôle of apoptosis in the pathogenesis of THEV is not clear, but this may be the cause of the immunosuppression.

Experimentally, the incubation period is five to six days following oral infection. In natural outbreaks, virtually ail birds become infected, as demonstrated by the development of antibodies. Mortality ranges from zero to over 60% with an average of 10%-15%.

Classically, the onset of disease is sudden. Birds are depressed, have bloody droppings and may die suddenly. Death usually occurs within 24 h of the appearance of the first signs of disease, or the bird recovers. Signs of disease within a flock last approximately six to ten days. Outbreaks due to less virulent strains are less spectacular. AU strains, including those previously thought to be apathogenic, are immunosuppressive. Therefore, infection with THEV may allow paramyxovirus type II, Chlamydia, Staphylococcus and E. coli to cause disease.

In a breeding organisation in Northern Ireland which has a very high standard of hygiène, breeding turkeys remained uninfected until commencement of lay. The turkeys developed a clinical condition similar to EDS, with loss of eggshell colour, thin shelled and shell-less eggs, and this was associated with seroconversion to THEV. The birds remained apparently healthy.

Birds which have died from the are often pale due to blood loss. Sudden death is often indicated by feed in the crop and good body condition of the carcass. The small intestine is usually distended, the mucosa is congested and the lumenfilled with feed and blood. In some cases, a yellow fibrinonecrotic membrane may be present. The lesions are more prominent in the proximal small intestine. If sick birds are killed, the spleens are found to be enlarged, friable and marbled or mottled. Where birds have died as a resuit of infection, the spleens tend to be smaller and the mottling is less apparent.

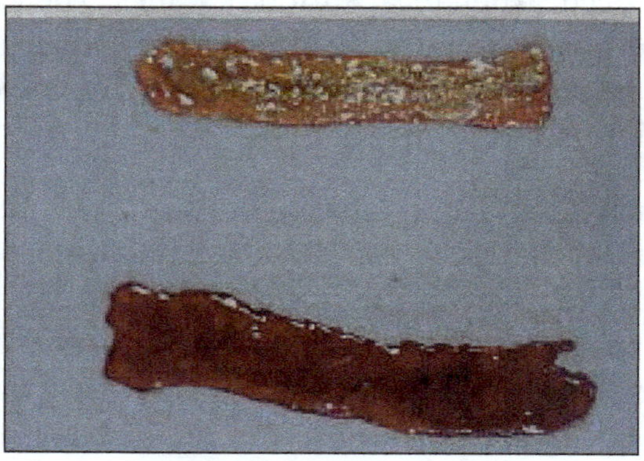

Duodenum from a turkey with haemorrhagic enteritis.
The upper specimen shows a less acute form.

Following experimental infection, proliferation of the white pulp surrounding the splenic ellipsoids occurs from day three onwards. This progresses to large irregular islands of white pulp, grossly visible around days five to six. At around three to four days, lymphoblasts, probably B cells, with intranuclear inclusions are prominent. Inclusions are also present in the splenic mononuclear phagocytes. By day four or five, the white pulp begins to become necrotic, and by day six to seven the white pulp is completely involuted with only occasional plasma cells appearing in the red pulp. Lymphoid depletion also occurs in the thymus and bursa of Fabricius between days three and nine.

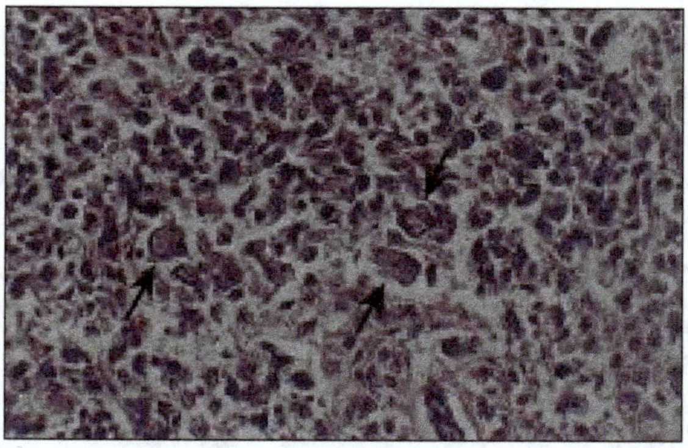

Spleen from a turkey which died as a resuit of infection with turkey haemorrhagic enteritis virus.
Several cells with large intranuclear inclusions are present (arrows) Haemotoxylin and eosin.

Severe congestion of the intestinal mucosa, degeneration and sloughing of the villous epithelium and haemorrhages in the villous tips are also observed. One group reported that the blood vessels in the lamina propria were intact and the erythrocytes appeared to escape from the vessels by diapedesis. Increased numbers of lymphoreticular cells with intranuclear inclusions are present in the lamina propria, in addition to mast cells, plasma cells and heterophils.

Diagnosis

The spleen is the preferred organ for virus isolation, but faeces also contain large amounts of virus. The lymphoblastoid B-cell Une of turkeys (MSTC-RP19) is inoculated. If cell culture is not available, then five- to ten-week-old antibody-free turkeys can be given material orally or by the intravenous route. Birds usually die approximately three days after the intravenous injection and five or six days after oral infection. Birds which are infected but still alive at six days usually have enlarged spleens.

Traditionally, diagnosis has been made using affected spleen as antigen in a double immunodiffusion test. More sensitive tests such as the immunofluorescent test, ELISA, restriction endonuclease and PCR are now being used increasingly.

Antibody can be first detected three to four days after infection using the ELISA. This antibody is long lasting; in one flock, 83% of the birds were still positive forty months after initial testing. Due to lower sensitivity, the double immunodiffusion test becomes positive only after two weeks. Further details are provided by Pierson.

Enlargement of the spleen in turkeys can be caused by THEV, but can also be due to reticuloendotheliosis or lymphoproliferative diseases. Blood in the intestine gives a strong indication of the, and demonstration of antigen in the spleen provides the proof.

Vaccines are used in many areas. A tissue culture attenuated vaccine has been used extensively, but such vaccines have been reported to be immunosuppressive. Vaccines derived from the spleen of birds with THE or MSD have been used as vaccines, but both types are also immunosuppressive. A recombinant fowl pox vaccine which afforded good protection under laboratory conditions and which did not cause immunosuppression has recently been reported.

Marble Spleen Disease

Marble spleen disease is observed in pheasant production operations throughout the world. Marble spleen disease occurs naturally in three- to eight-month-old birds, but has been experimentally reproduced in adult pheasants. Infection, as indicated by antibody development, approaches 100%. Birds are often found dead, but depression, weakness, nasal discharge and dyspnoea may also be observed. Mortality ranges from 2% to 20%, usually occurring over a period often to fourteen days, but can continue for several weeks.

Antigen is present in spleen, liver, lung, bone marrow and kidney, but in contrast to THE infection, no antigen is detectable in the intestine. Bursectomy protects against the disease and an age-related resistance occurs below six weeks of age, which is unconnected with the presence of maternai antibody. This indicates the importance of B lymphocytes in the disease process. T lymphocytes are important in controlling MSD infection. Infection with MSD impairs bodi the humoral and cell mediated responses. The effect on the humoral response is more pronounced and lasts several weeks.

The spleens of pheasants which have died as a resuit of MSD are usually enlarged and mottled or marbled. In naturally occurring cases of the disease, the lungs are congested and oedematous (which is thought to be the cause of death through asphyxiation). In contrast to HEV, no evidence exists of intestinal bleeding or lesions. The splenic weight is significantly elevated between 6 dpi and 10 dpi. Histological changes in the spleen are similar to those in THE. Splenic necrosis and numerous large intranuclear inclusions are usually observed in birds which have died from the disease. Necrosis may also be observed in the lungs.

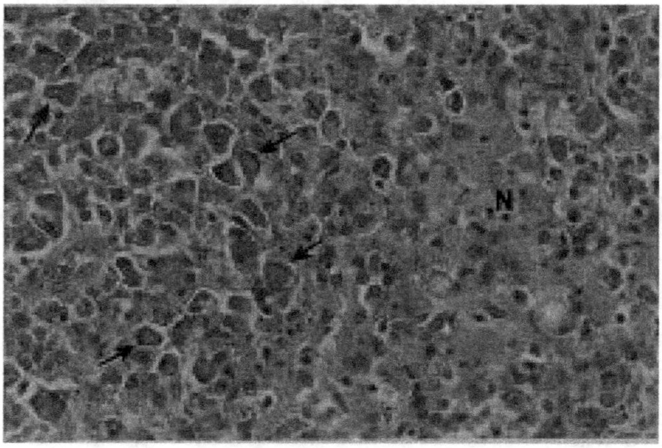

Spleen from a pheasant which died from marble spleen disease. Necrosis (N) and numerous cells containing large intranuclear inclusions are visible Haemotoxylin and eosin.

Avian Adenovirus Group II Splenomegaly

Antibody to avian adenovirus group II splenomegaly is widespread, but associated disease is not a major problem. Mortality is unusual, although 8%-9% mortality has been recorded in mature chickens. Infection is recognised as splenomegaly in broilers at slaughter and as splenomegaly with pulmonary oedema/congestion in adults. The disease can be important as a cause of condemnation at slaughter because of enlarged spleens.

Public Health Importance

The group II adenoviruses pose no threat to public health, as no record exists of infection of mammals by these viruses.

Control

The viruses are widely distributed throughout the world and therefore import restrictions would not be justified.

Group III Adenoviruses (Egg Drop Syndrome Viruses)

Aetiology

The EDS virus (EDSV) is a typical adenovirus, except that it agglutinates erythrocytes of chickens, turkeys, ducks, geese, pigeons and peafowl. The virus does not agglutinate erythrocytes from a wide range of mammals. It partially shares an antigen with FI adenoviruses.

Only one serotype of EDSV has been recognised, but three genotypes have been recognised using restriction analysis. One genotype encompasses isolates obtained from chickens in Europe over a period of eleven years. The second encompasses isolates from ducks in the United Kingdom and the third from chickens in Australia suffering from EDS.

Egg drop syndrome virus grows to high titres in duck kidney, duck embryo liver or duck embryo fibroblast cultures and chick embryo liver cells. The virus grows less well in chick kidney cells and grows poorly in chick embryo fibroblast cultures. Growth in turkey cells is poor and no growth could be detected in a range of mammalian cells. The virus grows to high titres in a range of goose cells.

In chick liver cells, peak virus and intracellular haemaggrutinin titres are reached approximately 48 h after infection, and peak extracellular haemagglutinin titres at approximately 72 h.

The virus grows very well in SPF embryonated duck or goose eggs and this is the best System for producing antigen for vaccine or haemagglutinins, as titres of 1/16,000-1/32,000 are produced.

Epidemiology and Pathogenesis

The behaviour of the EDSV in chickens appears unique compared to other adenoviruses. After initial entry through the nasal or gastrointestinal mucosa, local viral replication is followed by a transient viraemia. The principal site of virus replication is the pouch shell gland, and replication occurs to a lesser degree elsewhere in the reproductive tract. If the embryo is infected, or the chick is infected before sexual maturity, then the virus remains latent until the bird is sexually mature. This strategy ensures transmission of the virus to the next generation, as virus will be present in and on eggs for up to three weeks. Virus is excreted through the cloaca and originates in the oviduct. Unlike other adenoviruses, EDSV does not originate from the gastrointestinal tract, as the virus has minimal replication in this organ.

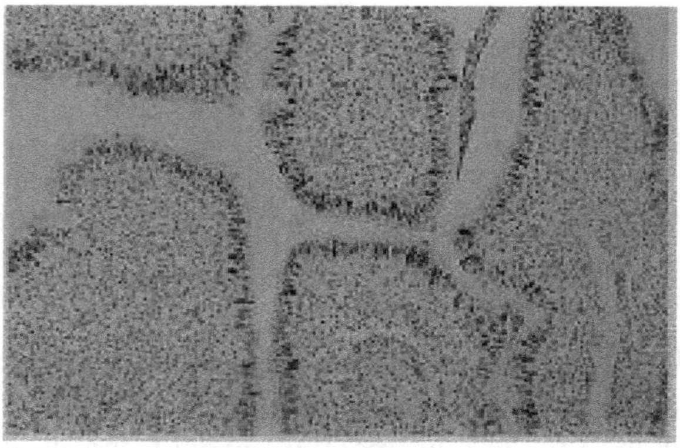

Shell gland from a hen infected with egg drop syndrome virus.
Viral deoxyribonucleic acid (brown) is abundant in the surface epithelium.
In situ hybridisation, haematoxylin counterstain.

AU ages and breeds of chickens are susceptible, although differences in the response may occur. When two brown egg laying strains and one white egg laying strain were experimentally infected, the white egg strain showed a more marked depression in egg production than the brown egg strains. However, the brown egg strains produced more eggs with shell defects.

Information on the pathogenesis of EDSV in waterfowl is scarce, but the available evidence suggests that EDSV behaves as a conventional adenovirus. Waterfowl are frequently infected with EDSV. Thus, studies on birds in the Atlantic flyways in the USA demonstrated antibody in ruddy (Oxyura jamaicensis), ring-necked (Aythya collaris), bufflehead (Bucephala albeold), wood (Abc sponsa), lesser scaup (Aythya affinis), mallard (Anas platyrhynchos), northem shoveler (Anas clypeata), gadwall (Anas strepera) and common merganser (Mergus merganser) ducks. Antibody has also been detected in coot (Fulica spp.), grèbes, cattle egrets (Bubulcus ibis), Canada geese (Branta canadensis), herring gulls (Larùs argentatus), owls, a stork and a swan. Antibody is also widespread in domestic ducks, including Muscovy ducks and geese Quail are susceptible and develop classical clinical signs. Although turkeys and pheasants can be experimentally infected, no signs of disease are observed. Guinea-fowl may be naturally infected and develop typical signs. However, in one study, guinea-fowl failed to show signs of disease after being infected with a fowl isolate.

Three syndromes are associated with EDS. The classical form was seen when primary breeding stock became infected. Chicks derived from thèse flocks remained healthy and did not produce antibody until reaching sexual maturity. At some time between the onset of egg laying and peak production, abnormal eggs were produced and the birds produced antibody. This infection probably initially arose from the use of a vaccine grown in duck cells which contained latent EDSV (18). Infection has since been eradicated from primary chicken breeding stock (18). However, the viras subsequently infected commercial egg-producing flocks and has become endemie in some areas. This is primarily due to the presence of virus on the exterior of eggs, leading to contamination of trays and trolleys. In many cases, this equipment is not adequately cleaned or disinfected before being returned from the egg packing plants to other faims at random. Infection can also be transmitted from flock to flock by humans, such as group advisory staff and workers servicing equipment.

The third category is the sporadic outbreak. This occurs when chickens corne into contact with domestic or wild waterfowl. Contact may be direct or through contaminated drinking water. Thèse outbreaks are self-limiting unless infection is spread to other flocks, when the outbreaks become the focus of an endemie cluster.

Clinical Signs

The first sign is loss of shell colour in pigmented eggs. This is quickly followed by the appearance of thin shelled, soft shelled or shell-less eggs. The thin shelled eggs often have a rough sandpaper-like appearance or a granular roughness at one end. If the obviously affected eggs are removed, fertility and hatchability are not affected. Small eggs have been described in some outbreaks, but were not reported in experimental infections. Some workers have described watery albumen, but others found normal albumen. Egg drop syndrome is probably a misnomer as much of the apparent drop is due to the production of shell-less eggs, as the numbers of shell membranes in the litter will testify, and consumption of the thin-shelled and shell-less eggs by birds.

If the disease develops as a resuit of reactivation of latent virus then the production of usable eggs is reduced by approximately 40%; this usually occurs when the flock is coming into lay and production is between 50% and peak production. Production does not return to predicted levels until four to ten weeks later and compensation often occurs later in lay resulting in an overall loss of eggs of ten to sixteen eggs per bird.

Range of egg changes observed as a result of infection with egg drop syndrome virus. Pale and thin-shelled eggs are shown in the top row.

S: Shell-less egg.

M: Ruptured shell membrane.

N: Normal pigmented egg.

If the birds become infected by lateral spread when in lay, the clinical picture may appear different because transmission can be very slow, especially if the birds are in cages. Poor egg production may be reported rather than a marked decline. A careful inspection will reveal that only birds in a few cages are producing abnormal eggs at any one time. The speed of spread is influenced by a number of factors, such as the number of birds initially infected and the position of the affected cages with respect to the flow of the belts transporting the eggs, feed and faeces.

Affected birds appear healthy; some workers have described inappetence and dullness but most have not reported this. Transient diarrhoea has been described by some, but this is probably exudate and fluid excretion from the oviduct.

Although one study reported oral infection of day-old chicks causing increased mortality in the first week of life, most workers have found no signs of disease either in chicks or growing birds infected with virus, or commercially in the very large numbers of chicks hatching from infected eggs.

Diagnosis

Selection of Specimens

Selection of the correct specimen is very important. If birds are vertically infected, then no antibody will develop until sexual maturity. Therefore, to certify that a breeding flock is free of vertically transmitted virus, testing should be undertaken no earlier than thirty-two weeks of age. In the absence of clinical signs, selection of the correct bird for sampling poses a problem. In a cage unit, this problem can be overcome by selecting a cage in which affected eggs are present. If ail the birds in the cage are examined, then at least some birds will have antibody and may have virus. The pouch shell gland is the organ of choice for histology, immunochemistry or virus isolation, but pathognomonic lesions and viral antigen are present only for a short time. If blood is to be collected for serology, then the birds bled should be those from the cages in which defective eggs have been produced for the longest time. On litter, the problem is more difficult. To isolate virus, or to detect antigen or lesions, the simplest method is to feed affected eggs to antibody-free hens held individually in cages. The eggs produced by these birds should be examined daily and testing should be performed when a bird produces abnormal eggs. Examination of randomly selected cloacal swabs has been successful in some cases.

Serological Tests

The haemagglutination inhibition (HI) test is the method of choice. A 1/10 serum dilution is mixed with an equal volume of a solution containing four haemagglutinating units of antigen. The mixture is allowed to react for 15 min at room temperature and then one volume of an 0.8% fowl erythrocyte suspension is added. Other tests, such as the ELISA and serum neutralisation, are available, but the HI test is rapid, inexpensive and accurate.

Virus Isolation

A 10% suspension is made from the pouch shell gland and the supernatant inoculated onto cell cultures or embryonated duck eggs. Suitable cells, in order of preference, are duck cells, chick embryo liver or chick kidney cells. At least fourteen days incubation (one blind passage) are required after inoculation. If the cells degenerate, the supernatant should be checked for the presence of haemagglutinins using a 0.8% fowl erythrocyte suspension. If agglutination occurs, the isolate can be confirmed by an HI test using specific antiserum.

Antigen Detection

Antigen can be detected in the pouch shell gland, during the time that defective eggs are produced, using immunofluorescent techniques on frozen sections or the avidin-biotin-peroxidase technique on formalin fixed tissue sections. In situ hybridisation may also be used.

Public Health Importance

Infection with EDSV has no public health significance.

Control

Basic breeding stock should be free of infection, and many breeding organisations are free at all levels. Given that EDSV is transmitted vertically and that birds do not develop antibody until sexual maturity, certification of freedom from vertically transmitted virus is not possible until the flock has been in lay for a number of weeks. The HI test is satisfactory, but the appropriate time for blood testing varies according to the type of parent. For broiler breeders, sampling at approximately thirty weeks of age would be acceptable, although thirty-five weeks was chosen in eradication programmes to allow a generous safety margin. Given the severe economic effects of the disease and the difficulty of excluding the virus if using an egg packing station serving infected flocks, many commercial egg producers routinely vaccinate flocks using a commercial inactivated vaccine which is very effective in controlling disease when administered correctly. Apparent failure of vaccines to protect appears to be due to poor vaccination techniques.

Risks from Imported Eggs and Processed Chickens

Since avian adenoviruses may be vertically transmitted, imported hatching eggs could give rise to infected chicks. Although adenoviruses occur world-wide, diseases such as hydropericardium syndrome do not. Therefore prudent measures would include checking the history of the supply flock, and rejection of eggs from a region or organisation where serious adenoviral diseases are present. Similarly, non-fertile eggs and hatchery waste eggs may also be infected, and these should not be recycled into poultry food.

Viraemia usually occurs in the early stages of adenovirus infections, and since adenovirus may be found in many visceral organs, adenoviruses could theoretically be present in processed chicken. However, flocks infected with significant adenoviruses will show evidence of disease and accordingly should not be slaughtered for human consumption. Therefore, the risk of importing very pathogenic adenoviruses with processed chicken should be low. Adenoviruses have been recovered from the faeces of normal chickens and the potential exists for contamination of carcasses in the processing plant. However, while adenoviruses may remain viable for some time, in contrast to some significant bacterial contaminants, the virus will not multiply on the carcass. Thus, while the risk of acquiring significant infection from uncooked poultry meat or offal appears small, care should nonetheless be taken to ensure that such poultry meat or offal are not recycled to avian species.

Egg Drop Syndrome Virus

Egg drop syndrome (EDS) is caused by a viral infection in laying hens. It is characterised by production of soft-shelled and shell-less eggs in apparently healthy birds, and leads to a sudden drop (10-40%) in recorded egg production or a failure to achieve a normal peak in production. It can be difficult to identify the early stages of the disease as hens will eat the shell-less eggs, and the only evidence that may remain is the membranes, which is a sign that is easy to miss. In flocks where some birds have acquired immunity due to the spread of the virus, a failure to reach expected production targets is observed. Clinical signs include diarrhoea and brief loss of shell colour and egg yolk pigment prior to the production of soft-shelled eggs and mortality is usually negligible. Ducks and geese are the natural hosts for the EDS virus and are asymptomatic carriers. Chickens of all ages and breeds are susceptible but the disease is most severe in broiler breeders and brown-egg layer strains. EDS was first introduced into chickens through contaminated vaccine and spread through breeder flocks. EDS is a notifiable disease in some states of Australia.

EDS can be distinguished from Newcastle disease and influenza virus infections by the absence of illness, and from infectious bronchitis by the eggshell changes that occur at or just before the drop in egg production.

What Causes Egg Drop Syndrome?

EDS is caused by infection with the EDS virus which is an adenovirus. The incubation period is three to five days and the course of the disease is four to 10 weeks. The virus is transmitted through any of the conventional means of viral disease spread and is also transmitted on and in the egg (horizontal and vertical transmission). The main method of horizontal spread is through contaminated egg trays, however, droppings are also infective. Contact with wild ducks or geese, or water or ranges frequented by these birds, may be a source of infection. Humans and contaminated fomites (such as crates or trucks) can spread virus, which can also be transmitted by needles when vaccinating and drawing blood. Insect transmission has not been proven but is considered possible. Chicks hatched from infected eggs may develop antibodies and excrete virus however it is more common that the virus will remain latent (alive but inactive) and subsequently the bird will not produce antibodies. The virus reactivates and grows in the oviduct when the hens go into lay, at which point the viral cycle begins again. Birds which are immune to the virus (already have antibodies) reduce the rate of spread of the virus. EDS has no effect on fertility or hatchability of eggs that are suitable for setting.

Treatment and Prevention of EDS

There is no successful treatment of EDS. The classical form has been eradicated from primary breeders and the maintenance of EDS-free breeding stock is the main control measure. In layers, induced moulting will restore egg production after an episode of EDS infection. The prevention of horizontal spread relies on good biosecurity and washing and disinfecting plastic egg trays before use can control the endemic form. The sporadic form can be prevented by ensuring that chicken flocks do not come into contact with other birds, especially waterfowl. As such, potentially contaminated water should be chlorinated before use and general sanitary precautions should always be followed. There are vaccines available to prevent infection, and if appropriately produced and administered, these inactivated vaccines can prevent clinical disease but will not prevent virus shedding.

5

Papillomaviridae and Polyomaviridae

Papillomaviridae and Polyomaviridae families have distinct biological features, but have similarities in virion structure, mechanism of replication and cell cycle regulation. Canine papillomavirus, papillomavirus of birds, polyomavirus, bovine polyovirus infection, etc. are some of its examples. All these diverse papillomaviridae and polymaviridae viruses have been carefully analyzed in this chapter.

Papillomaviridae

The Papillomaviridae is a family of small, non-enveloped viruses with double stranded DNA genomes of 5,748 bp to 8,607 bp. Their classification is based on pairwise nucleotide sequence identity across the L1 open reading frame. Members of the Papillomaviridae primarily infect mucosal and keratinized epithelia, and have been isolated from fish, reptiles, birds and mammals. Despite a long co-evolutionary history with their hosts, some papillomaviruses are pathogens of their natural host species.

Table: *Papillomaviridae*. Characteristics of the family *Papillomaviridae*.

Characteristic	Description
Typical member	Human papillomavirus 16 (K02718), species *Alphapapillomavirus 9*, genus *Alphapapillomavirus*, subfamily*Firstpapillomavirinae*.
Virion	Non-enveloped, 55 nm, icosahedral.
Genome	Circular dsDNA. Genome varies from 5,748 bp to 8,607 bp.
Replication	Bidirectional theta replication.
Translation	Early and late transcripts, alternative splicing, alternative open reading frames.
Host Range	Mammals, reptiles, birds, and fish.
Taxonomy	Two subfamilies include >50 genera and >130 species.

Virion

Morphology

Papillomavirus virions are non-enveloped. The viral capsid is ~600 Å in diameter and consists of 360 copies (arranged as 72 pentamers) of the major capsid protein, L1, and ~12 molecules of the L2 minor capsid protein (*Papillomaviridae*. Expression of recombinant L1 with or without L2 allows for self-assembly of virus-like particles (VLPs). Each capsid packages a single copy of the viral circular dsDNA. The packaged viral DNA is associated with core histone proteins.

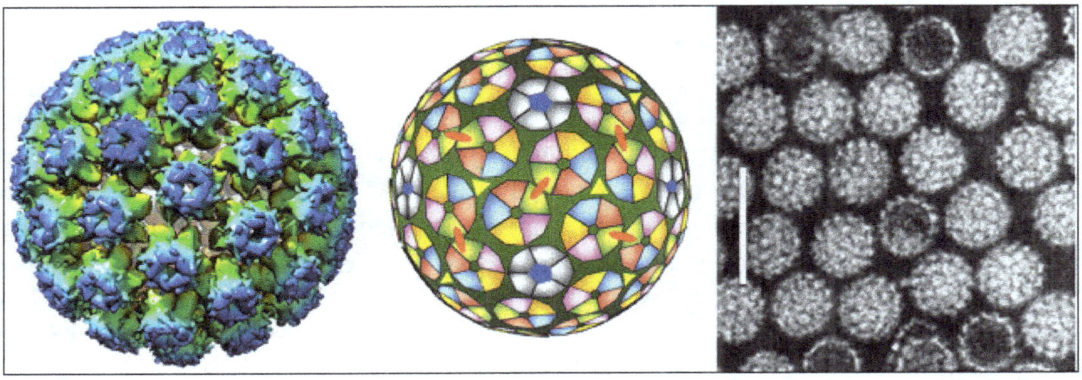

Papillomaviridae: (Left) Atomic rendering of a papillomavirus capsid. Derived from an image reconstruction from Cryo-electron microscopy of human papillomavirus type 16 at 4.5 Å resolution and colored according to the radial coloring scheme shown (PDB: 5KEP). (Center) Schematic diagram representing the 72 capsomers in a T=7 arrangement of a papillomavirus capsid. The icosahedral structure is composed solely of pentameric capsomeres for a total of 360 capsid proteins. (Right) Negative-contrast electron micrograph of human papillomavirus 1 (HPV1) virions. The bar represents 100 nm.

Physicochemical and Physical Properties

Nucleic Acid

The papillomavirus genome is approximately 7,500 bp. Viral genomes vary from 5,748 bp (Sparus aurata papillomavirus type 1; SaPV1) to 8,607 bp (canine papillomavirus type 1; CPV1). The genomes have an average GC content of about 42% (36–59%).

Proteins

Papillomavirus gene expression is tightly regulated at the level of transcription and RNA processing, including alternative mRNA polyadenylation and splicing. A typical papillomavirus encodes six to nine proteins. However, the ancestral papillomavirus may have only contained a core set of four proteins (E1, E2, L1, and L2). Temporal expression of the viral genome is associated with tissue differentiation. The viral DNA helicase (E1) is the only viral enzyme, and is essential for replication and amplification of the viral chromosome in the nucleus of infected cells. The viral E2 protein is the master regulator of the viral life cycle, and plays key roles in transcriptional regulation, initiation of DNA replication and partitioning the viral genome. The E1^E4 gene product is typically translated from a spliced mRNA fusing approximately the first four E1 codons to the E4 ORF, present in an alternative reading frame to the E2 ORF. A subset of viral mRNA encodes a short, hydrophobic, transmembrane protein, E5. Other than hydrophobicity, there is low sequence similarity among different E5 proteins. The E5 proteins can be further divided into different classes based on phylogeny and hydrophobic profiles. E5 proteins are typically encoded in the 3′-end of the early coding region. However, hydrophobic proteins, located in other parts of the viral genome have also been described. Designated as E10, these non-E5 hydrophobic proteins either overprint the E6 ORF, or are located in this region in the absence of an E6 gene. The productive phase of the viral life cycle occurs in differentiated cells that have exited the cell cycle. In order to complete the viral life cycle, the virus needs to uncouple replication from differentiation. The E6 and E7 proteins have been shown to play key roles in usurping the cellular environment to allow for replication. The E6 protein contains two zinc-binding motifs essential for its function. The E7 amino terminus contains regions of similarity to conserved regions (CR) 1 and CR2 of the mastadenovirus E1A protein, and the polyomavirus large T antigen. The E7 C-terminus contains a single zinc-binding motif homologous to the E6 motifs. The E6 and E7 proteins appear to be essential for members of the genus *Alphapapillomavirus*. Remarkably, the E6 and E7 proteins are not encoded by all papillomaviruses. The E8 exon is embedded within E1, and utilizes the same splice acceptor site as E1^E4 mRNA, generating mRNA for the E8^E2 protein. The E8^E2 viral repressor protein is present in essentially every known papillomavirus. E8^E2 inhibits viral replication and gene expression. Upon cellular differentiation, the viral capsid proteins L1 and L2 are expressed. The major capsid protein L1 is the structural component of the viral capsid. The minor capsid protein L2 plays an active role in viral assembly and throughout the infectious process.

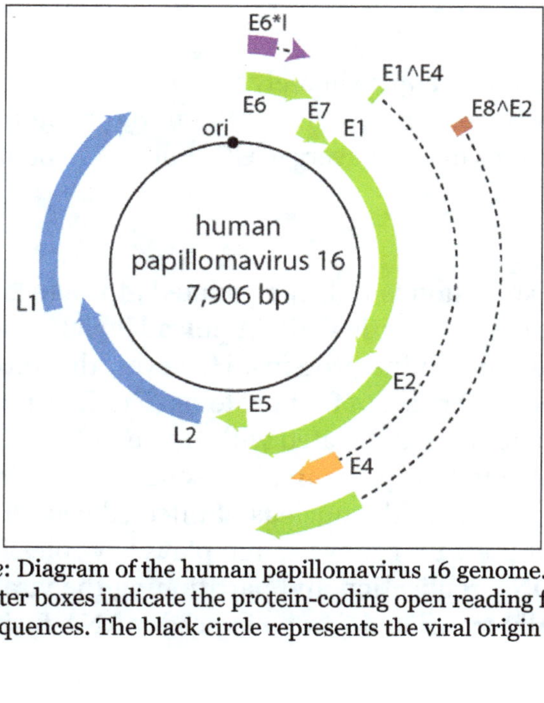

Papillomaviridae: Diagram of the human papillomavirus 16 genome. The viral dsDNA is indicated. The outer boxes indicate the protein-coding open reading frames. Dotted lines represent intron sequences. The black circle represents the viral origin of replication (ori).

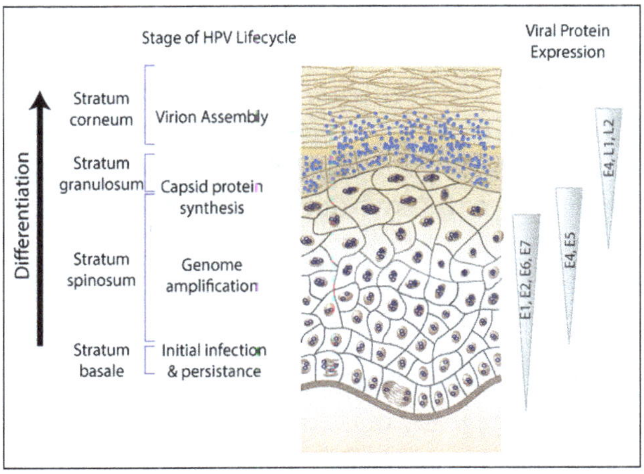

Papillomaviridae: Organization of the viral lifecycle. The different layers of the epithelium are shown on the left. The timing of expression and associated protein levels are summarized using triangles. Viral genome maintenance is facilitated by the expression of E6 and E7 together with E1 and E2. Increased levels of the viral replication proteins facilitate viral genome amplification. The expression of L1 and L2 allows for the formation of infectious virions (virus assembly).

Genome Organization and Replication

Transcription of the circular virus genome occurs from only one DNA strand. The viral genome can be divided into three functional regions. The early region encodes viral proteins involved in transcription, replication, and manipulation of the cellular milieu. The late region encodes the capsid proteins L1, and L2. The upstream regulatory region (URR or LCR) is located between the L1 and E6 ORFs and contains the viral origin of replication as well as binding sites for viral and cellular transcription factors.

The viral replication cycle consists of three distinct phases of replication. Initial limited viral DNA amplification is supported by the viral E1 and E2 replication proteins. The viral E2 protein binds to its binding sites in the viral origin of replication, recruiting the viral E1 helicase allowing for replication. This initial burst of replication is followed by maintenance replication, during which the viral genome is maintained at a relatively low, but constant copy number in the proliferating cells of a clonally expanded population of infected cells. Finally, as an infected cell completes cellular differentiation, there is a switch towards differentiation-dependent genome amplification and eventual generation of progeny virions. During maintenance replication, the virus needs to establish an S-phase like state in differentiated cells. Through a plethora of protein-protein interactions, the viral E6 and E7 proteins usurp the cellular environment, allowing for viral replication in differentiated cells. Remarkably, recent work has highlighted that different genera may induce this pseudo-S-phase through different mechanisms. The maintenance phase of the viral life cycle can last for months to years. In addition to regulating replication, the viral E2 protein plays a key role during maintenance by ensuring that the viral genomes are faithfully partitioned into the daughter cells. In the top layers of the differentiated epithelia, the viral DNA is amplified to a high copy number. The vegetative phase of the viral life-cycle requires the cellular DNA Damage Response. The viral capsid proteins self-assemble into particles encapsidating the viral DNA. As the cells slough off into the environment, infectious virions are released, completing the viral life cycle.

Antigenicity

Antigenicity is primarily determined by the major capsid protein, L1. Following vaccination, neutralizing epitopes typically map to a single variable loop, or more commonly two non-contiguous loops. Viral immunity appears to be highly species-specific, and there is only limited cross-protection, even to types within the same viral species. Furthermore, following natural infection, only approximately half of women seroconvert within 18 months following exposure. Prophylactic vaccines induce high-titer neutralizing antibodies restricted to a subset of (oncogenic) types within the genus *Alphapapillomavirus*. The current vaccines do not protect against types belonging to different genera. Vaccination with L2 (minor capsid protein) derived vaccines induces low-titer, yet broadly cross-neutralizing antibodies to heterologous PV types. These vaccines provide cross-protection in animal challenge models. Efforts to broaden the human papillomavirus (HPV) vaccines using L2 (poly)-peptides are currently underway.

Biology

Epidemiological and biological data is primarily available for the viral types belonging to the genus *Alphapapillomavirus*, and specifically those viruses associated with (cervical) cancer. An estimated 79 million Americans are infected, with an additional 14 million new HPV infections occurring every year. HPV is spread by skin-to-skin contact, and infections with genital human papillomaviruses are the most common sexually transmitted infection (STI).

Papillomaviruses primarily infect epithelial cells. Following a micro-abrasion, the incoming virion complexes with extracellular heparin sulfate proteoglycans on the basement membrane. This interaction results in conformational changes in the L1 and L2 capsid proteins, in turn allowing for the transfer of the virion to an unknown entry receptor. Following furin cleavage of L2, the virion becomes internalized using a process that shares similarities with macropinocytosis. Early trafficking events involve transporting virions from early endosomes into acidic late endosome and multivesicular bodies. This allows for capsid disassembly and uncoating. During this process, the viral DNA is believed to remain bound to L2. The L2-DNA complex traffics to the trans-Golgi network, remaining there until the onset of mitosis. During mitosis, the trans-Golgi network naturally vesiculates, and the vesicle-bound viral DNA finds its way into the nucleus. By metaphase, the viral DNA is associated with host chromosomes. Following mitosis, the viral DNA can be seen to be associated with nuclear ND10 bodies.

As the life cycle is completed in cells already destined for cell death, papillomaviruses are not viremic and are hidden from the immune system. In addition, papillomaviruses have evolved a plethora of mechanisms actively limiting the interferon response, a key antiviral defense mechanism. Overall, papillomaviruses appear to effectively evade the innate immune response, thereby delaying the activation of adaptive immunity. In turn, this likely plays an important role in persistence of the virus for months or even years.

Pathogenicity

Healthy skin harbors a large spectrum of different papillomavirus types belonging to different genera. While the majority of viral infections are subclinical, certain viral types cause (cervical, anal and oral pharyngeal) cancers, and have been associated with an increasing number of squamous cell carcinomas at specific sites. Based on their tropism, papillomaviruses can be roughly divided into cutaneous or mucocutaneous. Epidemiologically, the mucocutaneous HPV types can be further subdivided based on whether they are associated with benign or malignant lesions. Importantly, even in the case of viral types associated with specific pathologies, the majority of infections still present as subclinical. In the cervical environment, approximately 90% of HPV infections are cleared within two years post infection. Where clearance depends on an effective cell-mediated immune response, it is not clear why some infections are able to persist. Importantly, in the case of the oncogenic alphapapillomavirus types, persistent infection, not an incident infection is the main risk factor for progression towards cancer. Indeed, cellular transformation and viral replication are mutually exclusive, suggesting that oncogenic progression is not a typical outcome of infection.

Subfamily Demarcation Criteria

The current taxonomic classification of papillomaviruses is based on the nucleotide sequence of the L1 ORF. The L1 ORF of members of different subfamilies shares less than 45% sequence identity.

Genus Demarcation Criteria

The original criteria distinguishing genera stated: "Most types within a PV genus show less than 60% sequence identity to types of other genera based on global multiple sequence or pairwise alignments of the L1 genes. Nevertheless, the suggested percentage identities that define PV genera have to be taken as general, but not absolute criteria as curation is necessary". Practically, papillomavirus genera are primarily delineated by visual inspection of phylogenetic trees derived from concatenated E1, E2, L1, and L2 nucleotide sequences. Efforts are underway to refine the papillomavirus classification scheme.

Derivation of Names

Papilloma: from Latin *papilla*, "nipple, pustule", and suffix *-oma*, used to form nouns denoting "tumors". Viral genera belonging to the subfamily of the *Firstpapillomavirinae* are named according to the alphabet (e.g., *Alphapapillomavirus*). The prefixes "*Dyo-*"and "*Treis-*" are used to accommodate the growing list of viral genera within this subfamily. The names *Alphapapillomavirus*, *Betapapillomavirus*, and *Gammapapillomavirus* have been used for the genera containing papillomaviruses that infect humans and are not used in combination with the *Dyo-* or *Treis-* prefixes. Genera within the *Secondpapillomavirinae* are named according to the Semitic abjads. Currently, the *Secondpapillomavirinae* contain a single genus, bearing the first letter of this alphabet (transcribed as "A" in Latin): *Alefpapillomavirus*.

Phylogenetic Relationships

Phylogenetic analysis of papillomaviruses based on the concatenated alignment of four coding sequences (E1, E2, L2, and L1) from isolates from the type species of each of the 53 genera (*Papillomaviridae*) supports the existence of at least two distinct subfamilies (*Firstpapillomavirinae* and *Secondpapillomavirinae*). Likewise, this phylogeny, and that of an analysis of all 133 species, corroborates many genera and species within the *Firstpapillomavirinae*. However, not all genera or species are equally supported. There may be a need in the near future to base the taxonomy of the *Papillomaviridae* on the phylogenetic tree.

Similarity with other Taxa

There is evidence of recombination between a polyomavirus and a papillomavirus. The unclassified bandicoot papillomatosis carcinomatosis virus types 1 and 2 (BPCV1 and BPCV2) have circular dsDNA genomes encoding large and small T antigens related to avian polyomaviruses and capsid proteins (L1 and L2) of a putative marsupial papillomavirus.

Canine Papillomavirus

Canine papillomavirus can cause a range of skin problems in dogs, typically warty lesions (papillomas) that go away over time without specific treatment. Usually, a small to moderate number of warts are found in the mouths of infected dogs. Sometimes, the lesions can be large and extensive, causing problems and requiring surgical removal. Most often (like warts in people), it's merely a cosmetic issue that goes away eventually, but it may take months. Depending on the size and extent of the lesions, and whether they are growing or causing problems, treatment may be recommended. This can include crushing a wart to try to stimulate the body's immune response, or topical administration of certain drugs. Complete removal of the lesions surgically should be curative.

Canine papillomavirus is spread by direct contact between an infected dog and a susceptible dog. However, because it tends not to cause severe disease, this virus doesn't get a lot of attention in the research world. However, a recent study has provided more insight into the virus. In the study, researchers evaluated different ways to identify the virus. They then tested 95 healthy dogs at the University of Zurich and identified viral DNA on the skin or in the mouth of over 50% of the dogs. Whether these healthy dogs pose a risk to others is currently unclear, but it suggests that the virus could be transmitted not only from dogs with skin lesions, but from a large number of normal, healthy dogs. If that's the case, control will be a lot harder, since focusing on just the "sick" dogs (the ones with skin lesions) would miss a big pool of potentially infectious animals.

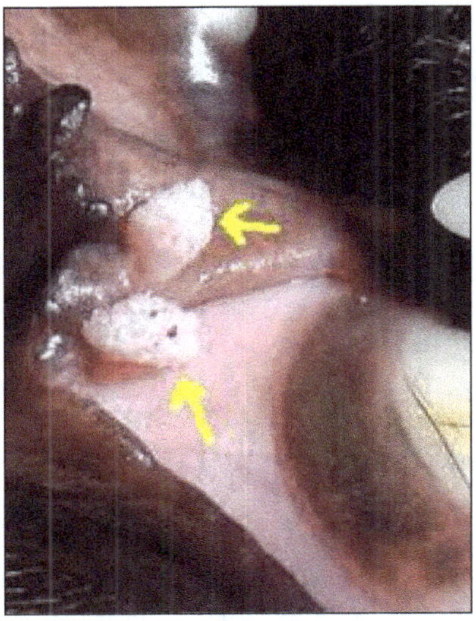

It's important to note that people cannot be infected with canine papillomavirus, nor can dogs be infected by papillomaviruses from other species.

Papillomavirus of Birds

Papillomas, or warts, are lesions or tumors that form on the skin of a bird which can take months to develop, may remain for years and then just resolve on their own. They are commonly found on the feet and legs or beak and head and have been diagnosed frequently in finches, waterfowl, cranes, herons, and flamingoes as well as African grey parrots and canaries. There are also reported cases of suspicious lesions occurring on various types of psittacine birds as well.

A papilloma is defined as a small, solid tumor (or lesion) having a clear-cut margin or border that is higher than the surrounding tissue of the skin. This lesion or tumor could have a peduncle (stalk) or appear more like a wart with various shapes.

Symptoms of Papillomas in Birds

Papillomas in birds can be external as well as internal, and while the external ones can be easily noted, the internal ones are not as likely to become apparent. Here are the symptoms you might expect to see:

- Lesions or warts noted on the skin of various parts of the body of a bird and are external (most likely locations will be on legs, feet, head and beak).
- Papilloma prolapsed from cloacal – internal.
- Internal papillomas will not be detectable by simple observation (those located in the gastrointestinal system which are caused by several herpesviruses and present predominantly in the Psittacine family).

Types

There are a couple of types of papillomas in birds, both benign and malignant:

- Papillomas of the skin, also known as "warts" - Usually benign and viral in nature, may disappear spontaneously or may need surgical intervention if they are large, irritated and bothersome.
- Squamous papillomas - Similar to warts or tumors noted above but are not viral in nature (tumor or wart is malignant).

Causes of Papillomas in Birds

Papillomas (also known as warts) are generally caused by a virus called the papillomavirus of which its family has many members. There are a number of quite specific viruses within this family which have been linked to papilloma development and these viruses are generally species specific in their inhabitation. These viruses can live outside the host's body for quite some time and, accordingly, are just waiting to infect a bird, being able to get into the skin of a bird via abrasions and cuts, through skin which has been softened by moisture and from insect bites (ticks, fleas, mosquitoes). They are not known to infect humans or unrelated species as they are considered very "host-specific".

Since the cause is viral, it is important to stress that the virus is contagious to other birds within the afflicted species. Papillomas are generally not a fatal disease to birds unless the bird has been infected with the Psittacine herpesvirus (PHV) which has been reported in the psittacine family and causes a deadly disease called papillomatosis. The virus which causes papillomatosis in the Psittacine family is similar to the papillomavirus that causes the less invasive and less dangerous papillomas in other bird species.

Diagnosis of Papillomas in Birds

If the lesions that you are noting on the skin of a bird are truly papillomas or warts, there is generally no diagnosis required. They will frequently resolve on their own as the host develops an immunity to the offending virus. If diagnosis is required or desired, your veterinary professional will need to do a physical examination, obtain a history from you and he can have the lesions tested by sending samples (called a biopsy) of the lesions to the laboratory for microscopic review. This type of analysis will likely only be necessary if the lesion or papilloma is large enough to become a problem for the bird when it eats, moves or stands.

Treatment of Papillomas in Birds

Treatment is generally not needed unless the papilloma is large enough or become irritated and bleeds. When treatment is needed, here are some things which could be recommended by a vet:

- General excision of the papilloma through the means of chemical cauterization or surgical removal works adequately.
- Isolation of the afflicted bird and any other birds who may also have been exposed will likely be recommended by a vet.
- Thorough cleaning and disinfection of the enclosure, food and water bowls, non-porous toys, to decrease the opportunity for the virus to spread.
- For those items which are porous and can't be thoroughly cleaned and disinfected (like wood, rope, natural fibers), it will be necessary to discard them to prevent continued re-infection of a birds.
- You should also monitor the fecal output on a daily basis in an attempt to make sure appropriate food consumption and digestion are taking place.
- Monitor the weight of the afflicted birds on a daily basis.

Recovery of Papillomas in Birds

Most birds recover from papillomas as long as they aren't internal ones caused by the Psittacine herpesvirus which can be much more dangerous to those birds in that family. Here are some suggestions and recommendations for what to do at home:

- Make every attempt to keep a pet bird from biting, scratching or licking to allow the inflammation to quiet down.
- Keep any ulcerative, inflamed, irritated or raw and bleeding papilloma clean to decrease the opportunity for bacterial or other types of infection.

- In terms of prevention, keep a pet bird separated from other birds, thus inhibiting the spread of any virus being carried by other birds who may not be sick.
- If you decide to increase an aviary, take the extra precaution to quarantine the new bird for a period of at least 90 days.
- Be sure to have a veterinary professional examine the new bird at the beginning and again at the end of the quarantine period to assure it is healthy enough to introduce to other birds.
- When you take one of the birds out of an aviary, be sure to quarantine it when it returns before re-introducing it to the remaining aviary population.

Budgerigar Fledging Disease

Budgerigars are relatively robust creatures, but can still fall victim to a wide range of ailments.

Budgies Avian Gastric Yeast (AGY) Infection

Also known as macrorhabdiosis, or megabacteriosis, this highly contagious infection is frustratingly difficult to spot in the early stages. The AGY incubates and proliferates in the bird with no outward sign of trouble. The first thing one might notice is that one's budgie loses weight, in spite of eating with his usual gusto. This is because the AGY impedes digestion. One may then spot undigested food in his droppings, or the bird may vomit food and mucus. He will also become listless.

Until 2004 the cause of the disease was thought to be a bacteria, but it has now been identified as a yeast, Latin name *Macrorhabdus ornithogaster*. The misdiagnosis came about due to bacteria, including Streptococcus, taking advantage of the budgie's hammered immune system and spreading secondary infection. The combination of the yeast and bacteria attack leads to a condition called wasting disease (sometimes called "going light").

A vet will be able to administer a drug to combat the AGY, and will recommend a healthy diet to aid recovery. This usually involves omitting all yeast-feeding sugary foodstuffs.

One will still need to keep a close eye on one's birds, as AGY has the horrible habit of lying fallow and then blooming again a few weeks later.

Budgie Candidiasis

This is another yeast infection. Candida, the organism responsible, is a form of thrush (the virus, not the bird!), and can bloom anywhere in a budgie's digestive system from the crop downwards. Some of the symptoms are similar to AGY infection – listlessness, vomiting and loose droppings. The vomit will have a nasty smell, and the bird's crop may swell up, due to gases produced by the Candida yeast. In advanced cases the budgie will suffer loss of balance and shaking fits.

Candidiasis can only be cured with drugs that kill the bacteria, so a trip to the vet is essential. The cure takes about one week, during which the budgie's diet should be closely controlled to avoid the ingestion of yeast-feeding sugars.

Budgie Sour Crop

This condition has more than one cause, but all the suspects are members of the yeast family. The symptoms are a swollen crop and sour-smelling vomit. Once again, it will take a targeted drug to kill the infection.

Candida lurks in the budgie's digestive system.

Budgie Sneezing

Sneezing, or coughing, is a symptom of a cold or similar virus in the budgie's upper respiratory tract. There will be an accompanying runny nose, or the cere will be caked in dried nasal discharge. A budgie sneeze is unlike a human's, but it's a noise one won't have heard him make before, and that will alert one to the fact that something is wrong.

The sneezing/coughing will be accompanied by other symptoms such as listlessness, panting, and spending a lot of time on the bottom of the cage. The bird may be short of breath, and may sometimes grip the sides of the cage with his feet and beak, stretching the neck in an effort to get more oxygen in. His breathing may be noisy: it will sound like a finger being rubbed on glass.

Various bacteria are involved in these budgie colds, and unlike a human cold they will not simply disappear after a few days. One can use a tissue on a tamed bird to soak up some of the nasal discharge, but one still need to get the bird to the vet as soon as possible.

Parrot Fever (Psittacosis)

This is the bird-borne disease most people have heard of, due to it being transmittable to humans. Chlamydophia psittaci is the organism responsible for the condition, and it is estimated that 1% of wild birds harbour the disease, a figure that rises to 30% in captive budgerigars. Most of these are carriers, showing no symptoms themselves, but passing the disease on via their droppings and saliva. Keeping cages clean is therefore the best way of preventing the disease from spreading.

The ill bird will need to visit a vet to verify his condition. Isolation is then vital, and the cage the bird came from will have to be disinfected. Monitoring other birds for symptoms, and remove any that one can think may have succumbed. The vet will prescribe drugs, or may recommend that it is in everyone's best interest to have the bird humanely killed.

Ill birds will need isolating.

Budgerigar Fledgling Disease

Also known as Papovavirus, this is caused by the Psittacine polyomavirus virus, which kills the young birds before fledging. It does not affect adult birds, although there is the possibility that they may be carriers of the virus. In a milder form, the virus produces a condition known as French moult.

Three week old budgie fledgling.

Budgie Wounds

Any wound on a bird can become infected and lead to septicaemia (blood poisoning). This will finish off a small bird like a budgie very quickly. If one see any traces of blood on the birds or in the cage, make a visual examination to spot the wounded party. It could be a case of feather bleeding, but is more likely to be an injury. Any flesh wound should be referred to a vet for antibiotic treatment.

Prevention is the best cure for this problem. Make sure the cage is free of sharp or pointed objects that could lead to wounds.

Budgie Splayed Feet

This is a condition that affects chicks who have been squatting on a flat nesting box floor while their leg bones develop. Setting up a nestbox correctly should prevent the problem from occurring. If the legs are splayed, the condition can be remedied by taping the legs together between the ankle and the knee for a few days, with enough slack for the bird to move around.

Polyomavirus

Polyomavirus is a deadly infection that affects many of the bird's body parts and organs simultaneously. This infection affects caged birds, especially those from the parrot family. Young birds from newborn to juveniles (14-56 days), are the birds most at risk and is usually fatal. Though not proven, adult birds are thought to form some immunity to polyomavirus.

Symptoms and Types

From the time the bird contracts the infection, it takes about 10-14 days for the it to display symptoms. However, a bird may or may not show any sign of the polyomavirus infection. If the symptoms are displayed in your bird, its death may be imminent usually within one or two days. Since the infection lowers the immunity of the bird, it can be susceptible to other viruses, bacteria, fungi and parasites, which can lead to secondary infection and death.

Birds with polyomavirus infection may display symptoms, including:

- A swollen (distended) abdomen,
- Loss of appetite,
- Regurgitation,
- Vomiting,
- Diarrhea,
- Dehydration,
- Weight loss,
- Depression,

- Feather abnormalities,
- Excessive urination,
- Difficulty breathing,
- Bleeding (hemorrhages) below the skin,
- Listlessness,
- Tremors,
- Paralysis.

Causes

The polyomavirus is usually contracted through direct contact with other infected birds. It is also contracted from infected feces, dander, air, nest boxes, incubators, feather dust or from an infected parent passing it to chick.

Treatment

There is no known treatment for the polyomavirus disease.

Prevention

Following strict hygiene methods, such as disinfecting nest boxes, cages, incubators or utensils, can help ensure your bird does not get infected with the polyomavirus. The virus, however, is resistant to most disinfectants; use oxidizers like chlorine beach instead. Aviaries and pet stores should also regularly screen for the virus. And new birds should be quarantined to ensure they do not carry the disease.

Vaccination is available, but its effectiveness has still not be proven. The vaccine is given as a double dose to young birds. The first dose is given at four weeks of age, and second dose is given between six to eight weeks of age.

Adult birds also receive a double dose of vaccination; the second dose given about two to four weeks after the first. A booster dose of the vaccine is then required yearly.

Bovine Infection

In evaluating the viral risk of bovine products several factors have to be evaluated. These include the risks associated with individual viruses, the probability of the viruses being present in the material and the procedures used to inactivate or clear contaminating viruses.

An important aspect of the risk assessment is that viruses weakly pathogenic in their host may be highly pathogenic in heterologous hosts. Ovine Herpes virus 2 is non-pathogenic in sheep, but causes high mortality when transmitted to cattle. Similarly Herpes-B, a herpesvirus of monkeys, causes minimal disease in its host species, but is a fatal infection in humans. Moreover, parenteral introduction of viruses in a therapeutic may overcome some of the barriers normally limiting cross species transmission. Several categories of viruses are of concern:

- Zoonotic viruses known to be transmissible from animals to humans. Amongst those viruses considered as zoonotic in cattle are: Rabies virus, Reovirus 3, Parainfluenza virus 3, rotaviruses, noroviruses, Cache Valley virus, West Nile virus, and tick borne encephalitis viruses.
- Animal viruses that replicate in human cells in vitro. Some animal viruses replicate in human cells in vitro but evidence of their zoonotic potential is weak or absent. For instance, the human population is seronegative for infection by Bluetongue virus, but the virus replicates efficiently in human cells and has been proposed as an oncolytic vector.
- Viruses that infect human cells, fail to undergo productive replication but may initiate disease. Experimental transmis-sion of certain members of the herpesvirus, adenovirus and polyomavirus families can result in tumour formation outside their natural host. These events are associated with abortive replication and often with integration of part of the viral genome into chromosomal DNA.
- Virus families that have shown a propensity to change host range. For instance, a single mutation in Feline parvovirus 1 may have been responsible for its initial transmission to the dog population, an event that is believed to have resulted in the death of 25% of the young puppies in the first phase of the outbreak. One of the most disturbing findings in recent years has been the discovery of new parvoviruses in bovine serum, whose properties and pathogenicity are only just beginning to be evaluated. Retroviruses are also included in this group because of the potential severity of the diseases associated with this family and their recent history of cross species trans-mission.

Probability of Contamination

Some zoonotic viruses are geographically restricted by vectors or other factors and pose a negligible risk to bovine materials harvested from the US, Europe, Australia and New Zealand. Nevertheless, each of these regions has their own viruses of concern, such as Ross River virus, a zoonotic alphavirus found in Australia. Within these regions, the probability of an endemic virus being in a particular material, like foetal bovine serum (FBS), is dependent on a number of factors; the frequency of infection, the ability of a virus to cross the placenta and establish a viraemia in the foetus, the length of the viraemic period and the size of the serum pool. Other products like collagen, thyroid hormones and trypsin require their own unique, but similar assessments.

Sporadic Virus Infections

Some virus infections, particularly those transmitted by arthropods, are only sporadic. This is well illustrated by Cache Valley virus contamination of US origin FBS. BioReliance have recorded four major episodes of fermenter contamination by this virus. While contaminations by this virus are uncommon they are very serious. Cache Valley virus is a zoonotic virus associated with fatal encephalitis. It grows to high titre in CHO fermenters and, therefore, poses a major biohazard and a formidable decontamination problem.

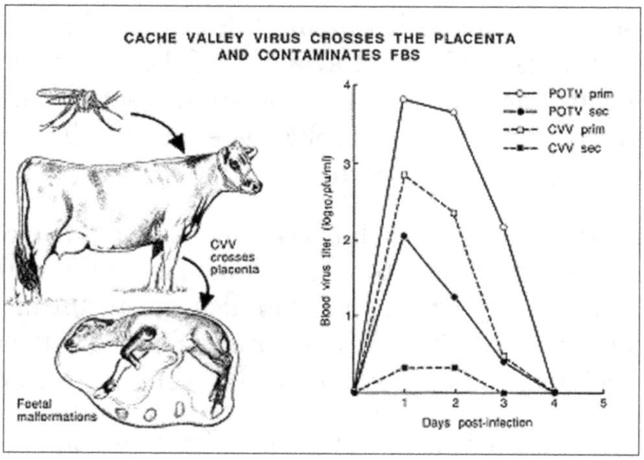

Contamination by this virus reveals the limitations of standard serum testing conducted by some suppliers of serum. Typically only 50ml to 100ml may be tested in an infectivity assay out of a commercial lot of 1000 litres. A pool of that size may contain samples from a thousand individual foetuses. Given that at peak viraemia there may be only 103 to 104 iu/ml, and for most of the short viraemic period the titre will be much lower, the final titre in the pool may be below the detection limit in a 50ml sample. This was confirmed by detailed analysis of one fermenter crash that occurred several days after initiation. No change in cell viability was noted until a few hours before the crash when all the cells died. The replication rate of the virus was determined and by back calculation it was shown that only 10 to 100 virions entered the fermenter in 20 litres of serum. Applying this very low level of virus input into our model predicted a crash within a few hours of the actual crash.

The recent contamination of Genzyme's manufacturing pro-cesses by Vesivirus 2117 is another example of a rare sporadic contaminant that caused significant economic loss. Only two examples of this calicivirus contaminating fermenters have been recorded. Vesiviruses are of concern because of their ability to replicate in the cells of a wide range of species and their known ability to jump species. One notable example was the development of a new disease in pigs, vesicular exanthema, following the feeding of contaminated sealion meat to pigs. The sealion virus spread from the US to Europe and also infected the human population, before expensive and rigorous procedures eradicated the disease.

The vesivirus that contaminated the commercial processes almost certainly entered in bovine serum. Antibody to vesiviruses in cattle is variable with some herds being sero-negative and others having high seropositivity; it is possible that cattle are sporadically infected from some other reservoir.

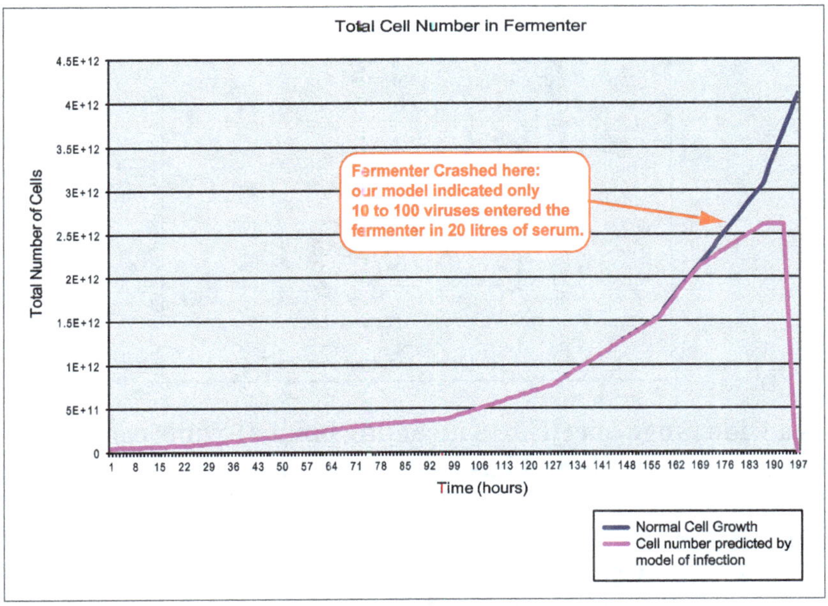

Modelled cell population dynamics in a fermenter following introduction of 1 to 100 infectious units of Cache Valley virus.

Common Virus Infections Bovine Viral Diarrhoea Virus (BVDV) and HoBi Virus

In contrast to Cache Valley virus and Vesivirus 2117, a very high proportion of FBS lots may contain Bovine viral diarrhoea virus (BVDV). The reasons for the high prevalence are related to the biology of the virus. Transplacental transmission of BVDV can lead to foetuses and calves that are immunotolerant and persistently viraemic. Estimates of foetal infection in unvaccinated herds vary from 1 in 100 to 1 in 1000 so that every serum pool is likely to contain one or more contaminated samples. The viruses present in foetuses are non-cytopathic and can go unnoticed as contaminants in cell cultures unless PCR or immuno-staining are employed to detect the virus. FBS lots are screened by serum producers using an infectivity assay that detects non-cytopathic virus but the inherent statistical uncertainty in sampling a large pool means that positives are missed. A solution is to institute sub-pool testing and rejection of positive sub-pools but this adds to material costs. For donor calf-serum, it has been possible to develop closed herds that are free of BVDV but again this is an expensive procedure. Given these uncertainties it is prudent for manufacturers to re-test the serum before use.

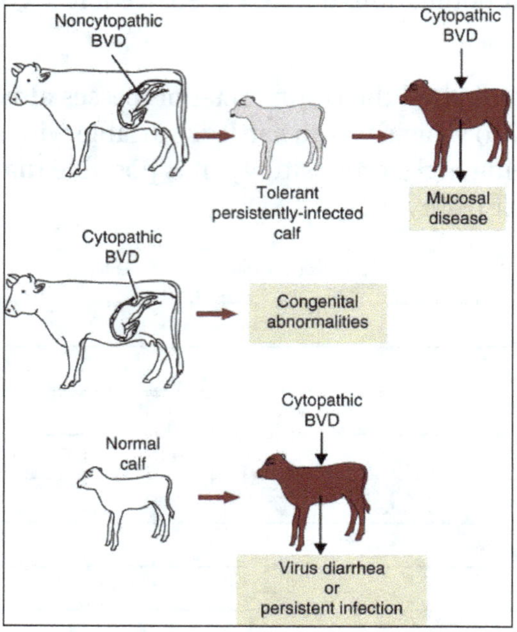

BVDV infects a wide range of cell lines including bovine, ovine, canine, feline, equine, lapine, simian and insect cells and has contaminated veterinary vaccines with devastating results. Caution is required in extrapolating species susceptibility from a few cell lines as in some cases resistant variants can be selected. Human cells have been regarded as resistant to BVDV but recently there has been a report of human cell infection. It is generally assumed that BVDV does not infect widely used production lines like CHO and NS0, however, recorded host ranges should never be taken as absolute, as even minor mutagenic change can profoundly alter host range and pathogenicity.

BVDV-1 & 2 are members of the genus Pestivirus within the family Flaviviridae, which also contains classical Swine fever virus, and Border disease virus. Recently a new virus HoBi, related to but distinct from, BVDV-1 & 2 has been isolated from contaminated cells. The source of the contamination was serum that had been screened for BVDV. HoBi virus, which may be renamed BVDV-3, should be detected in serological assays for BVDV using polyclonal antisera but several monoclonal antibodies used in BVDV diagnosis do not cross react with HoBi. Similarly, standard PCR tests for BVDV cells miss this virus and a specific PCR should be employed if cells banks are being screened.

In 2007 a new pestivirus, Bungowannah virus, was identified in Australian pigs with myocarditis. This virus appears to be an emerging infection and is currently restricted to Australia. It is known that pestiviruses can cross species barriers, for instance BVDV can infect pigs. As Australia is a major supplier of FBS, it is important that a watching brief is maintained for the presence of this virus in bovine serum.

BVDV enters the serum pool from immunotolerant, persistently infected foetuses. However, if a foetus is infected late in the third trimester it can mount an immune response and develop antibody to BVDV. This antibody can neutralise BVDV present in the pool and may mask infectious virus. Consequently in Europe, tests for "inhibiting antibody levels" need to be conducted before determining if infectious BVDV virions present. High levels of inhibiting antibody can result in rejection of the pool. There are some differences between the CVMP and CPMP requirements but these are encompassed in the tests described below. These mandated tests are based on the incorrect assumption that BVDV is monotypic for neutralisation whereas, in practice, the use of different BVDV strains can profoundly alter the result.

Bovine Polyomavirus (BPyV)

Bovine polyomavirus is another virus that is extremely common in virus pools when assessed by PCR and, in this case, there are significant concerns for human therapeutics. The virus had originally been discovered as a productive infection in primate cells and was initially thought to be a monkey virus until it was determined that it was a contaminant from serum used in cell culture. BPyV is of particular concern because, like other members of the polyomavirus family, it is can oncogenically transform cells in culture. There is good serological evidence that it is a zoonotic virus; Parry and Gardener demonstrated that seroconversion was common in people with occupational exposure to cattle (71% in veterinarians) but essentially absent in the general population. No detailed follow up molecular study has been conducted to determine whether BPyV sequences are found in cancers of at risk populations.

The high prevalence of the virus by PCR poses a particular problem as a PCR positive result does not necessarily indicate the presence of an infectious virus. This has led to the development of an infectivity assay that should be conducted on unirradiated serum where a PCR positive result is recorded. As discussed below, where the PCR signal is above a threshold it is also advisable to conduct infectivity assays on irradiated serum to ensure no infectious virus has survived. It should be noted that standard infectivity assays for bovine viruses do not detect BPyV and a specific assay involving multiple passages with PCR endpoints is required to detect low level contamination.

Recent Developments in Testing Raw Materials

Many of the assumptions about the frequency of particular viruses in serum have had to be radically revised following the introduction of massively parallel sequencing, sometimes referred to as deep sequencing. Massively parallel sequencing (MPS) is a powerful new method for the identification of viruses and other adventitious agents, without prior knowledge of the nature of the agent. Enormous amounts of data are developed in this process, usually between 100 to 400 Mbp. Consequently, sophisticated bioinformatic algorithms are required to analyse and verify virus targets.

The basis of the method is to optimise the isolation of putative viral sequences from serum or tissue while decreasing cellular genomic nucleic acids, to vastly reduce the complexity of the system. Typically this involves nuclease treatment to remove cellular sequences followed by nuclease inactivation and capsid dissociation in the presence of chaotropic agents and finally, recovery of undamaged, encapsidated nucleic acid. The remaining DNA and RNA targets are randomly primed and sequenced.

Allander first applied this approach to bovine serum resulting in the surprising discovery of two new bovine parvoviruses BPV-2 and BPV-3. Studies by BioReliance scientists using MP-Seq confirmed these findings and resulted in the finding of a new parvovirus BAAV-2, a member of the dependovirus genus. As discussed below, these are very frequent contaminants of serum and parvoviruses are amongst the most resistant viruses known, posing a challenge for inactivating procedures. Little is known of the tropism of BPV-2 & 3, even within their host species, but this family of viruses have shown major changes in host range. The onset of the canine parvovirus pandemic around 1979 is believed to have followed cross transmission of a feline virus following a single mutation in the capsid gene. In contrast, BAAV-2 and possibly the other bovine dependovirus BAAV-1, has a wide host range with BAAV-2 able to infect human cells.

New parvoviruses were not the only surprising discoveries. In a survey of four different FBS serum lots from major manufacturers, 2 out of 4 batches had complete sequences of bovine noroviruses and 2 also had sequences of kobuviruses. In both cases it was possible to reconstruct the complete genomes of these viruses and, as the samples had been nuclease treated, these genomes were contained within capsids and therefore potentially infectious.

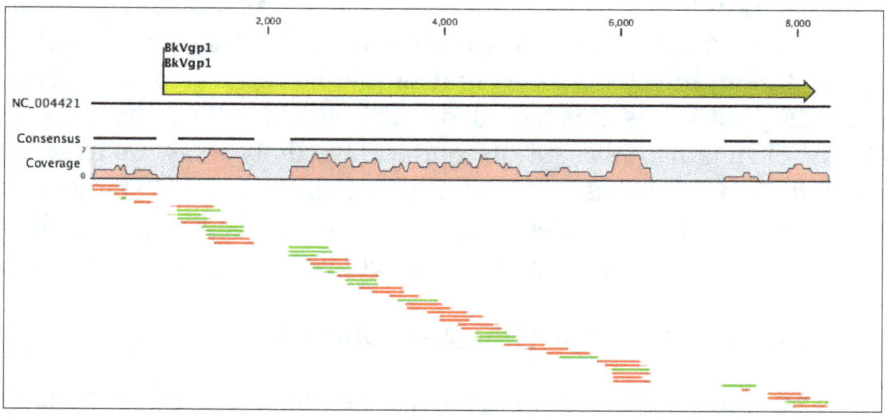

Detection of Bovine Kobuvirus by MP-Seq.

Noroviruses are known causes of diarrhoea in humans and there is evidence to suggest cross species transfer of bovine viruses to the human population. Kobuviruses are particularly interesting and important contaminants. This member of the Picornaviridae was unknown to veterinary science until it was detected as a contaminant of HeLa and Vero cells. Since then it has been recorded as an important cause of diarrhoea in young calves. The MP-SeqTM data regarding its presence in FBS suggests this is not a rare contaminant and it has the potential to be as serious a contamination agent as Vesivirus 2117 or Cache Valley virus. The virus can be cytolytic in cell cultures but until more is known about the virus, there is a case for adding specific PCR endpoints to in vitro cultures. This is supported by the data on the detection of vesivirus in fermenters, where the virus only became detectable towards the end of 14 days culture period.

Integrating Testing and Risk Mitigation with Quality by Design

As the biotechnology industry matures there is increasing em-phasis on Quality by Design (QbD) principles as formulated in ICH guidance document Q8 (R2) Pharmaceutical Development. Encompassed within QbD is a control strategy designed to ensure that a product of required quality will be produced con-sistently. Elements of the control strategy focus on input materials and the "design space" that affects control of those materials. The difference between the traditional approach and a QbD approach to raw materials is worth examining; FBS is used in the example below but the principles apply to all bovine materials.

Required, Traditional Screening Methods for FBS and Bovine Materials

The CPMP note for guidance on the use of bovine serum, and the European Pharmacopoeia recommend that serum producers test serum before inactivation. A combination of specific and general tests are used to detect: Bovine viral diarrhoea virus (BVDV), Bovine adenovirus (BAV), Bovine parvovirus (BPV), Bovine respiratory syncitial virus (BSRV), Reovirus type 3 (Reo 3), parainfluenza virus type 3, (PI3V), infectious bovine rhinotracheitis virus (BHV-1), Rabies virus (RV) and Bluetongue virus (BTV). Virus infection is determined by a combination of cytopathic effect, haemadsorbtion and specific immunofluorescence assays.

The US Code of Federal Regulations requires a similar approach using Vero cells and bovine cells. Bovine turbinate cells are employed because of their high susceptibility to BVDV. Although PI3V and BHV-1 are not specifically mentioned they are detectable in standard 9 CFR tests for bovine viruses. The tests are prescriptive in their requirements but it is possible to design protocols that meet both European and US regulatory requirements.

Title	Duration (weeks)	Assay description	CVMP assay	CPMP assay	Combined CVMP/CPMP assay	US assay
Determination of inhabitation levels of bovine serum on multiplication of BVDV.	6	4 passages, then 2 week titration	032940	032920	032931	

Determination of inhibition levels of bovine serum of serum on detection BVDV.	2	2 week titration	032941	03221	032932	
Detection of viral contaminants in bovine serum.	6	4 passages, 8 spot slides	032942	032922	032933	
Detection of BVDV in bovine serum.	6	4 passages, 2 chamber slides for BVDV	032943	032923	032934	
Combined CPMP/9 CFR.	6	4 passages, 2 chamber slides, plates	n/a	032930	n/a	032930
Bovine 9 CVR.	4	2 passages, 2 chamber slides, plates	032910	n/a	n/a	032900 (7 viruses) 032901 (9 viruses)

These in vitro assays achieve the aim of screening for the viruses that were considered of main concern over two decades ago. However as MP-SeqTM technology has shown, these are not necessarily the commonest virus infections and they specifically miss detecting critical viruses like bovine polyomaviruses. Vesivirus 2117 is also likely to missed in these assays as it replicates more efficiently in CHO cells than in standard bovine indicator cells (BioReliance unpublished). At a minimum it is now usual to add screening for bovine polyomavirus initially by PCR, followed by infectivity assays for positive material. Some manufacturers have also put in place specific screening for Cache Valley virus and Vesivirus 2117. One advantage of PCR in the latter cases is that the presence of non-infectious virus increases the sensitivity over an infectivity assay, in the case of Cache valley virus the PCR assay was about 400 times more sensitive.

A New Approach to Raw Material Quality Control

An inherent part of traditional testing strategies was the belief that it was possible, with a high degree of certainty, to select sera free of adventitious agents and that if a material passed a 9 CFR or CPMP test it was safe to use. The greater understanding of viruses present in serum that has come from new technologies like MP-Seq, emphasises the need for a quantitative risk based approach.

A new approach to raw material quality control involves 3 or 4 steps:

- Understanding the universe of potential contaminants in the raw material.
- Developing specific, quantitative assays for those viruses, taking account of the statistical limitations of sampling from the raw material pool.

- Relating the potential viral load in a given batch of raw materials to inactivating procedures like gamma-irradiation.
- Where no inactivating steps are in place for the raw material, adding monitoring assays later in the process to ensure the viruses are eliminated.

Understanding the range of contaminants that may be present is best determined through the use of new technology like MP-Seq that makes no assumptions about the nature of the virus (or other biological contaminant), or its ability to replicate in a set of pre-determined indicator cells. MP-Seq is not likely to become a routine batch by batch quality control tool until sequencing costs fall further. However, several manufactures are now embracing the concept of reviewing the data from MP-Seq on several batches of raw materials from a given supplier. This approach should be linked to agreements that tightly specify the geographical source of the materials so that the MP-Seq data are reflective of the universe of contaminants from that supply source.

This technology provided indications that new viruses like BPV-2, BPV-3, BAAV-2, Bovine norovirus and Bovine kobuvirus were frequent, and often high level, contaminants of serum. The next stage is to develop specific assays for these viruses. In the case of BPV-2 and 3 permissive cell systems have not been identified and therefore specific PCR assays have been used to determine the frequency and level of viral genomes in serum. These specific PCR tests confirmed the high frequency and the very high levels of circulating genomes present.

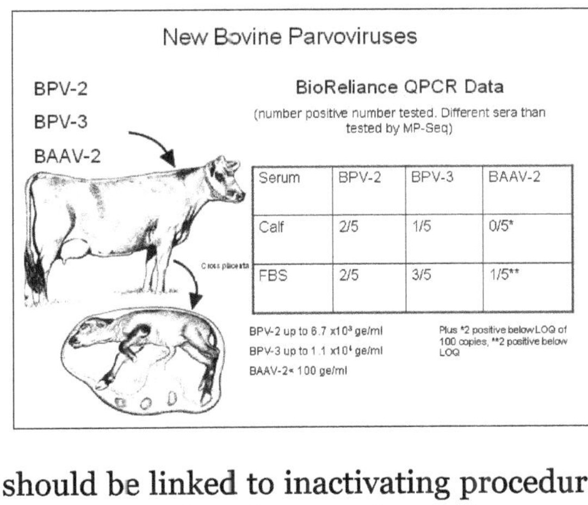

Finally the viral load should be linked to inactivating procedures. In Europe it is now a requirement to use gamma-irradiated serum in vaccine manufacture, but in a QbD approach it is important to understand the limitations of inactivation by irradiation. Standard irradiation involves treatment with 35Gy, but where batch irradiation is used, outer parts of the batch may receive higher doses impairing the quality of the serum. The kinetic inactivation curves for gamma irradiation are essentially first order. The dose required to produce a 1 log10 inactivation of the virus, or D value, varies between viruses but lies in the range of 3.9 to 5.3kGy for several major groups. Protection in a serum environment is likely to increase protection for viruses and, as Plavsic & Bolin demonstrated, ssDNA viruses like circoviruses and parvoviruses are remarkably resistant to irradiation. This has important consequences for analysing FBS which may contain BPV-2 and BPV-3 genomes at levels above the capacity of irradiation to inactivate. An appropriate approach is to screen batches by quantitative PCR using only those batches with a low level of genomes. For instance, the control might specify an inactivation capacity 3 log10 greater than the virus load.

Where no serum inactivating steps are in place then, as part of the QbD approach, appropriate in process tests should be conducted. An evaluation of the capacity of a downstream purification process to inactivate or remove the contaminants identified in serum should also be undertaken. Implementation of this approach would have avoided the catastrophic contamination of rotavirus vaccines by porcine circoviruses introduced in contaminated trypsin.

References

- Papillomaviridae, ictv-reports-ictv-online-report-dsdna-viruses: talk.ictvonline.org, Retrieved 02 June, 2019
- Canine-papillomavirus: wormsandgermsblog.com, Retrieved 18 July, 2019
- Papillomas, bird-condition: wagwalking.com, Retrieved 19 March, 2019
- Diseases, budgie-guide-health-problems: omlet.co.uk, Retrieved 14 February, 2019
- C-bd-Polyomavirus, bird-conditions-neurological: petmd.com, Retrieved 25 January, 2019
- Selection and characterization of canine, swine and rabbit cell lines resistant to bovine viral diarrhea virus. J Virol Methods. 2006 Oct;137(1):51-7. Epub 2006 Jul 18

6

Parvoviridae and Circoviridae

Parvoviridae virus comprises of sixty copies of a single protein sequence, called VP1, VP2 etc. Circoviridae virus is non-enveloped consisting icosahedral and round geometries. This chapter delves into the study of feline panleukopenia virus, canine parvovirus, goose parvovirus, porcine circovirus, chicken anemia virus, etc. to provide in-depth knowledge of the subject.

Parvoviruses

Parvovirus in dogs is a potentially fatal infection that can damage the intestines and result in severe diarrhea and dehydration. Canine parvovirus type 2 (CPV-2) emerged in late 1970s causing severe epizootics in kennels and dog shelters worldwide, and soon became endemic in the global dog population.

Feline panleukopenia virus has almost identical DNA sequence as canine parvovirus, but this one does not harm dogs. It causes disease in all felids, as well as in some members of related families (such as raccoons or minks). It is commonly referred to as feline distemper and causes fever, low white blood cell count, diarrhea and sometimes even death.

Porcine parvovirus is the major causative virus in a reproductive failure syndrome in swine, are characteristic symptoms are grouped under an acronym SMEDI (which stands for stillbirths, mummified fetuses, early embryonic death, and infertility). This infection has been reported to occur worldwide with variable prevalence rates.

Two specific parvoviruses of rats, RV (Kilham rat virus) and H-1 virus, are highly pathogenic for fetal and infant rats. Although the infections are usually subclinical, they can also seriously damage central nervous system, liver, lymphoid system, and other tissues. In colonies where the infection is endemic, a majority of rats develop antibodies and become immune by the time they are seven months old.

Feline Panleukopenia Virus

In the past, feline panleukopenia (FP) was a leading cause of death in cats. Today, it is an uncommon disease, due in large part to the availability and use of very effective vaccines. The disease is also called feline distemper or feline parvo.

Feline panleukopenia (FP) is a highly contagious viral disease of cats caused by the feline parvovirus. Kittens are most severely affected by the virus. The names feline distemper and feline parvo should not be confused with canine distemper or canine parvo — although their names are similar, they are caused by different viruses. The viruses do not infect people.

The feline parvovirus infects and kills cells that are rapidly growing and dividing, such as those in the bone marrow, intestines, and the developing fetus.

Cats Susceptible to FP

Because the FP virus is everywhere in the environment, virtually all kittens and cats are exposed to the virus at some point in their lives.

While cats of any age may be infected with the feline parvovirus that causes FP, young kittens, sick cats, and unvaccinated cats are most susceptible. It is most commonly seen in cats 3-5 months of age; death from FP is more common at this age.

The virus has appeared in all parts of the United States and most countries of the world. Kennels, pet shops, animal shelters, unvaccinated feral cat colonies, and other areas where groups of cats are housed together appear to be the main reservoirs of FP. During the warm months, urban areas are likely to see outbreaks of FP because cats are more likely to come in contact with other cats.

Infection on Cats

Cats can shed the virus in their urine, stool, and nasal secretions; infection occurs when susceptible cats come in contact with these secretions, or even the fleas from infected cats. An infected cat tends to shed the virus for a relatively short period of time (1-2 days), but the virus can survive for up to a year in the environment, so cats may become infected without ever coming into direct contact with an infected cat. Bedding, cages, food dishes, and the hands or clothing of people who handle the infected cat may harbor the virus and transmit it to other cats. It is, therefore, very important to isolate infected cats. Any materials used on or for infected cats should not be used or allowed to come in contact with other cats, and people handling infected cats should practice proper hygiene to prevent spreading the infection.

The virus that causes FP is difficult to destroy and resistant to many disinfectants. Ideally, unvaccinated cats should not be allowed into an area where an infected cat has been — even if the area has been disinfected.

Diagnosing FP

The signs of FP can vary and may be similar to other illnesses such as Salmonella or Campylobacter infection, pancreatitis, feline immunodeficiency virus (FIV) infection, or feline leukemia virus (FeLV) infection. Infected cats may even show signs that resemble those seen when a cat has been poisoned or has swallowed a foreign object.

The FP virus causes damage to the cells that line the intestines. It also attacks the bone marrow and lymph nodes, resulting in shortages of all types of white blood cells (panleukopenia) and of red blood cells (anemia). The first visible signs an owner might notice include generalized depression, loss of appetite, high fever, lethargy, vomiting, severe diarrhea, nasal discharge, and dehydration. Sick cats may sit for long periods of time in front of their water bowls but not drink much water. In some cats, the fever will come and go during the illness and abruptly fall to lower-than-normal levels shortly before death. In young kittens, the virus can also damage the brain and the eyes.

Pregnant female cats that are infected with the virus and become ill (even if they do not appear seriously ill) may abort or give birth to kittens with severe damage to the cerebellum, a part of the brain that coordinates nerves, muscles and bones to produce body movements. These kittens are born with a syndrome called feline cerebellar ataxia, and their movement is accompanied by severe tremors (shaking).

Feline panleukopenia may be suspected based on a history of exposure to an infected cat, lack of vaccination, and the visible signs of illness. When that history of exposure is combined with blood tests that show severely reduced levels of all white blood cell types, FP is likely the cause of the cat's illness. FP is confirmed when the feline parvovirus is found in the cat's stool, but the results might be falsely positive if the cat was vaccinated for FP within 5-12 days prior to the test.

Treatment

The likelihood of recovery from FP for infected kittens less than eight weeks old is poor. Older cats have a greater chance of survival if adequate treatment is provided early. Since there are no medications capable of killing the virus, intensive care and treatment are critical to support the cat's health with medications and fluids until its own body and immune system can fight off the virus. Without such supportive care, up to 90% of cats with FP may die.

Treatment focuses on correcting dehydration, providing nutrients, and preventing secondary infection. Although antibiotics do not kill the virus, they are often necessary because infected cats are at a higher risk of bacterial infections because their immune systems are not fully functioning (due to the decreased white blood cells) and because bacteria from the damaged gut may enter the cat's bloodstream and cause infection.

If the cat survives for five days, its chances for recovery are greatly improved. Strict isolation from other cats is necessary to prevent spread of the virus. Other cats that may have been in contact with the infected cat, or in contact with objects or people who were in close contact with the sick cat, should be carefully monitored for any visible signs of illness. In most cases, once a cat recovers from FP, it will not infect other cats through direct contact, but some recovered cats can shed the virus in their stool and urine for up to 6 weeks.

Prevention

Cats that survive an infection develop immunity that likely protects them for the rest of their lives. Mild cases that go unnoticed will also produce immunity from future infection.

It is also possible for kittens to receive temporary immunity through the transfer of antibodies in the colostrum — the first milk produced by the mother. This is called "passive immunity," and how long it protects the kittens from infection depends upon the levels of protective antibodies produced by the mother. It rarely lasts longer than 12 weeks.

Prevention is vital to your cat's health. Today, there are vaccines that offer the best protection from feline parvovirus infection. Vaccination is equally important for strictly indoor cats as well as indoor/outdoor cats because the virus is everywhere in the environment. Most young kittens receive their first vaccination between six and eight weeks of age and follow-up vaccines are given until the kitten is around 16 weeks of age. Adult vaccination schedules vary with the age and health of the cat, as well as the risk of FP in the area. Consult your veterinarian for advice on an appropriate vaccination schedule for your cat(s).

Canine Parvovirus

Canine parvovirus is a highly contagious virus that can affect all dogs, but unvaccinated dogs and puppies younger than four months old are the most at risk. Dogs that are ill from canine parvovirus infection are often said to have "parvo." The virus affects dogs' gastrointestinal tracts and is spread by direct dog-to-dog contact and contact with contaminated feces (stool), environments, or people. The virus can also contaminate kennel surfaces, food and water bowls, collars and leashes, and the hands and clothing of people who handle infected dogs. It is resistant to heat, cold, humidity, and drying, and can survive in the environment for long periods of time. Even trace amounts of feces from an infected dog may harbor the virus and infect other dogs that come into the infected environment. The virus is readily transmitted from place to place on the hair or feet of dogs or via contaminated cages, shoes, or other objects.

Tank, a parvovirvus survivor.

Signs of Parvovirus

Some of the signs of parvovirus include lethargy; loss of appetite; abdominal pain and bloating; fever or low body temperature (hypothermia); vomiting; and severe, often bloody, diarrhea. Persistent vomiting and diarrhea can cause rapid dehydration, and damage to the intestines and immune system can cause septic shock.

Most deaths from parvovirus occur within 48 to 72 hours following the onset of clinical signs.

Parvovirus infection is often suspected based on the dog's history, physical examination, and laboratory tests. Fecal testing can confirm the diagnosis.

No specific drug is available that will kill the virus in infected dogs, and treatment is intended to support the dog's body systems until the dog's immune system can fight off the viral infection. Treatment should be started immediately and consists primarily of intensive care efforts to combat dehydration by replacing electrolyte, protein and fluid losses, controlling vomiting and diarrhea, and preventing secondary infections. Sick dogs should be kept warm and receive good nursing care. When a dog develops parvo, treatment can be very expensive, and the dog may die despite aggressive treatment. Early recognition and aggressive treatment are very important in successful outcomes. With proper treatment, survival rates can approach 90%.

Since parvovirus is highly contagious, isolation of infected dogs is necessary to minimize spread of infection. Proper cleaning and disinfection of contaminated kennels and other areas where infected dogs are (or have been) housed is essential to control the spread of parvovirus. The virus is not easily killed, so consult your veterinarian for specific guidance on cleaning and disinfecting agents.

Preventing Parvovirus

Vaccination and good hygiene are critical components of prevention.

Young puppies are very susceptible to infection, particularly because the natural immunity provided in their mothers' milk may wear off before the puppies' own immune systems are mature enough to fight off infection. If a puppy is exposed to canine parvovirus during this gap in protection, it may become ill. An additional concern is that immunity provided by a mother's milk may interfere with an effective response to vaccination. This means even vaccinated puppies may occasionally be infected by parvovirus and develop disease. To reduce gaps in protection and provide the best protection against parvovirus during the first few months of life, a series of puppy vaccinations are administered. Puppies should receive a dose of canine parvovirus vaccine between 14 and 16 weeks of age, regardless of how many doses they received earlier, to develop adequate protection.

To protect their adult dogs, pet owners should be sure that their dog's parvovirus vaccination is up-to-date. There are titers available that measure the dog's level of antibodies against the canine parvovirus, but the antibody level may not directly translate to protection if the dog is exposed to the virus.

Until a puppy has received its complete series of vaccinations, pet owners should use caution when bringing their pet to places where young puppies congregate (e.g. pet shops, parks, puppy classes, obedience classes, doggy daycare, kennels, and grooming establishments). Reputable establishments and training programs reduce exposure risk by requiring vaccinations, health examinations, good hygiene, and isolation of ill puppies and dogs. Contact with known infected dogs and their premises should always be avoided.

In spite of proper vaccination, a small percentage of dogs do not develop protective immunity and remain susceptible to infection.

Finally, do not let your puppy or adult dog to come into contact with the fecal waste of other dogs while walking or playing outdoors. Prompt and proper disposal of waste material is always advisable as a way to limit spread of canine parvovirus infection as well as other diseases that can infect humans and animals.

Dogs with vomiting or diarrhea or other dogs which have been exposed to ill dogs should not be taken to kennels, show grounds, dog parks, or other areas where they will come into contact with other dogs. Similarly, unvaccinated dogs should not be exposed to ill dogs or those with unknown vaccination histories. People who are in contact with sick or exposed dogs should avoid handling of other dogs or at least wash their hands and change their clothes before doing so.

Mouse Parvovirus

There are two important parvoviruses of mice: minute virus of mice (MVM) and mouse parvovirus type-1 (MPV-1).

Minute virus of mice (MVM) is an important infectious agent in laboratory mice. It is a ssDNA virus of the family Parvoviridae. Multiple strains have been described.

Mouse parvovirus type 1 (MPV-1) is a recently recognized and important infectious agent in laboratory mice. It is a ssDNA virus of the family Parvoviridae and was formerly known as orphan parvovirus. Three isolates of one serotype have been identified.

Transmission

The parvoviruses require rapidly dividing cells (such as GI, skin, and lymphoid organs) to survive. They are shed in urine and feces and may be transmitted via respiratory routes. They are highly contagious and shed virus for an undetermined time after infection. Direct contact with affected animals is probably required, but transmission has not been well characterized.

It is important to realize that it is not known if intermittent shedding of the virus occurs when mice harboring virus in the intestines are exposed to environmental or experimental stressors.

MVM has been reported as a common contaminant of transplantable tumors and mouse leukemia virus stocks.

Clinical Signs

Mouse strains vary in their sysceptibility to MVM. Natural parvovirus infections of immunocompetent and immunocompromised mice generally result in no overt clinical disease or pathology. Viral replication occurs in the pancreas, small intestine, lymphoid organs, and liver that persists for several weeks after infection. MVM also replicates in the kidneys.

Experimental infection with MVM will result in severe damage to multiple organs in the developing fetus and in neonates.

Diagnosis

Diagnosis is usually based on serology, via ELISA or IFA or both. MVM and MPV-1 may cross-react during the IFA test due to similar nonstructural proteins in the viruses.

Effects on Research

The parvoviruses have similar effects on research. They can reduce the rate of transplantable tumor take by direct oncolysis or modulation of immune response to tumor cells, interfere with the selection of new transplantable tumor phenotypes, and cause a reduction in viral or chemical tumorigenesis. In immunological studies, they can interfere with the modulation of lymphocyte mitogenic responses, interfere with ascites production, result in cryptic infection of lymphocytes, and interfere with the humoral antibody spectrum. They can interfere with infectious disease studies and cell biology studies by modifying interviral interaction and causing immunosuppression of the host. They can also result in altered patterns of rejection of skin allografts. A newly identified rat parvovirus (RPV-1) may suppress lymphoid tumor development.

Prevention

To prevent this disease, obtain replacement stocks from sources that are known to be free of disease. Tumor lines should be assessed for infection using MAP tests or other appropriate tests. Personnel working with infected animals should not enter rooms that contain naïve animals.

Eradication

The most effective way to eradicate parvovirus infections is to cull the colony and obtain clean replacement stock. However, this is not always a feasible option when working with valuable mice.

Caesarian rederivation or embryo transfer can be used to produce offspring that have not been exposed to the virus. Repeated serological evaluations should be performed prior to reintroduction of the mice into a naïve population.

A breeding moratorium of at least 8 weeks can also be used to prevent the spread of the virus from young weanling animals to younger naïve animals. The animals should be housed in microisolator caging and handled with standard microisolator techniques. This method requires repeated serologic testing and strict adherence to a zero-tolerance for breeding policy. It is important to note that transgenic and knockout mice often have altered immune systems that may allow them to sustain the infection for longer periods of time or to develop a carrier state. In these cases, the breeding moratorium would not be the appropriate means of eradication.

Goose Parvovirus

Goose Parvovirus is a highly contagious condition of geese and young Muscovy ducks. The younger the bird affected the more acute the condition and the higher the mortality. Losses are negligible in birds over 5 weeks of age.

Derzsy's Disease is caused by a parvovirus distinct from chicken and mammalian parvoviruses. The amount of maternal antibody passed from the breeding birds will affect the severity and timing and severity of the condition in the young birds. Vertical transmission resulting in congenital infection may occur.

Signs

- Prostration and death in acutely affected goslings.
- Reduced feed intake.
- Excessive water intake.
- Swollen eyelids and eye and nasal discharge.
- Profuse white diarrhoea.
- Membrane covering tongue.
- Loss of down.

- Reddening of skin.

Diagnosis

Signs and lesions in birds of the appropriate age and species.

Treatment

No specific treatment. Antimicrobials may be of value in reducing the effects of secondary bacterial infections.

Prevention

Hatching and brooding geese from different parent flocks together should be avoided. Ideally flocks that have suffered the disease should not be used for breeding as they may become persistent excreters of the infection.

Administration of immune serum has been shown to be effective but may require two doses (day old and around 3 weeks). The preferred approach is to immunise breeding birds with an attenuated live vaccine.

Circovirus

Porcine circovirus (PCV) is a common virus of pigs found throughout the world. This DNA virus is unique for its small size (~1800 bp), circular genome, and hardiness in the environment. In the late 1990s, a novel circovirus emerged in North American swine that appeared to be genetically distinct from the prototype PCV (now referred to as PCV Type 1) and was termed PCV Type 2 (PCV2). Porcine circovirus Type 1 had been recognized for several decades as a common contaminant of laboratory cell cultures but despite being widespread in commercial swine, was nonpathogenic. The emergence of PCV2 coincided with the occurrence of a new clinical syndrome of swine referred to as postweaning multisystemic wasting syndrome (PMWS). Since its appearance, PCV2 (and PMWS) has become widely distributed in most developed swine industries around the world. The term PCVAD (porcine circovirus associated disease) is now used in North America to refer to the different disease manifestations associated with PCV2. In Europe, the acronym PCVD (porcine circovirus disease) is used instead.

Porcine circovirus Type 2 infection has been reported from nearly every country with a significant commercial production industry. It is strongly associated with the occurrence of PMWS and also appears to have an association with porcine dermatitis and nephropathy syndrome (PDNS), porcine respiratory disease complex (PRDC), and occasionally reproductive failure. Postweaning multisystemic wasting syndrome usually occurs in nursery or growing pigs and is diagnosed on a herd level through the occurrence of multiple features:

Microscopic	lymphocytic depletion in lymph nodes often accompanied by histiocytic infiltration
Clinical	loss of body condition in affected individuals, enlarged lymph nodes
Epidemiologic	increased proportion of wasting pigs, lack of response to therapy in affected individual accompanied by a high case fatality rate

Numbers of reported cases of PCVAD have increased rather dramatically in recent years. PCVAD now is widespread in the United States, Europe, and reported from most swine-producing areas. An apparent immune-mediated sequel to viral (perhaps PCV2) infection, porcine dermatopathy and nephropathy syndrome (PDNS) is sporadically reported. Circovirus is often associated with both PMWS and PDNS but PCV2 as sole cause remains in dispute. Other agents or risk factors are likely to be involved.

In utero infection with PCV2 may be inapparent, associated with reproductive failure or abortion. Because PCV2 is ubiquitous in swine populations and other agents or risk factors are frequently present, disease associations with PCV2 are sometimes difficult to prove.

Porcine circovirus Type 2 can contribute to PRDC in some herds. Experimental co-infection studies as well as anecdotal evidence suggests that PCV2 is likely to increase the severity of pneumonia caused *Mycoplasma hyopneumoniae*, swine influenza virus, porcine reproductive and respiratory syndrome virus (PRRS) and others.

Since 1994, outbreaks of PMWS have been reported in North America and Europe and most swine-rearing areas of the world. The outbreaks have been consistently associated with PCV2 infection. Continued research and characterization of PCV2 remains of major interest among swine practitioners, researchers and producers. Genomic sequencing and restriction fragment length polymorphism (RFLP) techniques have defined several distinct viral clades within the PCV2 genotype but the clinical relevance of these distinctions is not completely understood.

In 1982, a circovirus was isolated from porcine kidney cell line (PK-15) as an adventitious virus. Circoviruses are small, nonenveloped animal viruses with single-stranded circular DNA. Porcine circoviruses are quite hardy in the environment and in organic substrates. They are sensitive to most disinfectants but chlorhexidine, ethanol, and iodine are less effective. Circoviruses now are known to cause psittacine beak and feather disease, an infectious anemia in chickens, and lethal infections of pigeons, canaries and ¬finches. Two types of circoviruses in swine have been described. The PK-15 virus is referred to as Type I (PCV1). There is circumstantial evidence that PCV1 may be involved in congenital tremors but it is generally considered to be nonpathogenic for swine. Analysis has demonstrated that PCV associated with PCVAD in swine is distinctly different genetically and antigenically and it is now classified as PCV2. Antibodies between the two types have low cross-reactivity.

PCV2 is consistently present in pigs with PMWS, and disease has been reproduced by experimental infection with PCV2. Many but not all field cases of PCVAD have concurrent viral infections or some other evidence of immune stimulation that seems to allow permissive replication and pathologic effects of PCV2 in swine. Herds that are endemically infected with other significant pathogens, are operated on a continuous flow basis, are in swine dense areas, provide a suboptimal environment (air quality, hygiene), or practice poor biosecurity have been shown to be more likely to experience an outbreak of PCVAD. Once affected, these same herds often are presented with a more severe form of the disease.

Epidemiology

The epidemiology of porcine circovirus infections is speculative. Nearly all US swine herds are seropositive. Seroconversion usually occurs by two to four months of age irrespective of whether clinical signs of PCVAD are observed.

With PMWS, affected pigs are between two and four months of age and "waste away" while cohorts seem to grow normally. Morbidity varies from 2-30% but case fatality is high, approaching 80%. Transmission of the virus and conditions allowing virulence expression are being vigorously studied. Some risk factors purported include coinfection with other viruses (PRRSV, porcine parvovirus, etc.) or nonspecific immune stimulation (vaccination). Activated macrophages appear to be permissive for viral replication and persistence.

Pathogenesis

Pigs infected with PCV2 develop viremia of variable duration, with replication and persistence most extensive in macrophages and monocytes of lymphoid organs and lung. Lymphoid depletion at multiple sites, chronic lymphohistiocytic to granulomatous inflammation, and an erosive bronchiolitis with fibrosis are characteristic. Reproduction of typical lesions by inoculation with PCV2 has been inconsistent but there is general agreement among researchers that the virus is at least necessary and perhaps causal in order for PCVAD to be expressed. Mechanisms that allow disease expression and immunity are not definitively known but appear to be related to macrophage activation prior to infection with PCV2. Occasionally, pigs may develop blotchy purple skin lesions and nephropathy, likely as an immune mediated sequel to viral infection, which is termed porcine dermatopathy and nephropathy syndrome (PDNS).

Clinical Signs

PCVAD: Clinical signs include gradual wasting, unthriftiness, rough hair coat, polypnea, dyspnea, pallor, diarrhea, and occasionally icterus. To be defined as PCVAD, there must also be histologic lesions of lymphoid depletion and lymphohistiocytic to granulomatous inflammation in affected organs with PCV2 demonstrable within lesions.

Affected pigs usually die; clinical survivors are severely stunted; non-clinical pigs in the same groups perform quite well. Both signs and lesions, as reported by various investigators, vary considerably.

PDNS: Is mainly a condition of pigs from 8-18 weeks of age. There are red-purple blotches on the skin, sometimes slightly raised, most obvious on the hind legs and perineum but can extend over the abdomen eventually covering the whole body. Most pigs with PDNS eventually die.

Lesions

PCVAD: Gross lesions include pallor, marked enlargement of all lymph nodes and perhaps spleen, and interstitial pattern of pneumonia in lungs. Less often present are icterus, a fluid-filled intestine, subcapsular white foci on the kidneys, atrophy of the liver and ulcers of the pars esophagea of the stomach. Microscopy reveals depletion of germinal centers in lymphoid tissues with replacement by histiocytes and multinucleated giant cells. Basophilic intracytoplasmic inclusion bodies often can be seen in macrophages in lymph nodes, tonsils, spleen and Peyer's patches. Lymphohistiocytic to granulomatous inflammation can be present in any organ, typically lungs and lymphoid tissues, and PCV2 antigen is abundantly present within characteristic lesions by immunohistochemistry (IHC) or *In situ* hybridization (ISH). Erosive bronchiolitis and fibrosis of airways is fairly typical.

PDNS: Often, there are small circumscribed circular to irregular, deep-purple discolorations of skin that become apparent on hind quarters, perineum, and flanks. Over the course of several days, the skin discolorations either resolve or expand and coalesce. Necropsy often reveals kidneys to be swollen with foci of pallor and hemorrhage in parenchyma. Microscopically, there is nonsuppurative vasculitis in sections of skin. Kidneys often have swollen glomerular tufts, inflammation, and protein in tubules. Multifocal nonsuppurative interstitial nephritis can also be a feature. The lesions are thought to be due to immune mediated (immune complex) vasculitis.

Lesions of PDNS are not specific for porcine circovirus.

Diagnosis

PCVAD: Diagnosis is made on clinical features, gross lesions, microscopic lesions, and demonstration of PCV2 within lesions. Using microscopy, characteristic lesions may be observed and virus inclusion bodies can often be identified in depleted lymph nodes, spleen, tonsils or Peyer's patches. Immunohistochemical tests or in situ hybridization are used to confirm diagnosis of disease associated with PCV2. Polymerase chain reaction (PCR) will confirm the presence of virus but does not confirm the presence of disease. Similarly, isolation and identification of a circovirus is not useful for diagnosis since asymptomatic infection is widespread in swine populations. Serological testing for both types of PCV has been developed to confirm infection but finding antibodies to circovirus in the herd is not diagnostic since most herds are positive without discernible disease. PCVAD must be differentiated from other bacterial or viral diseases, especially PRRSV which has similarities in clinical signs and lesion. Concurrent infections are common.

PDNS: Diagnosis of PDNS is by observing typical clinical signs, gross lesions, and histopathology. PCV2 antigen is demonstrable in some but not all cases.

Reproductive failure: In cases where reproductive failure is observed as abortions, stillbirths, and mummification, PCV2 can be demonstrated by IHC in fetal hearts as well as other tissues, in a routine diagnostic investigation.

Control

There is no specific treatment for pigs with PCVAD. However, anti-inflammatory agents and antimicrobials are sometimes used with some success in groups of pigs experiencing disease. All in/all out pig flow, thorough cleaning and disinfection between batches of pigs and early segregation and euthanasia of affected pigs are measures that may help control the disease. Antibiotics in feed or water may suppress secondary infections. Better definition of risk factors and intervention strategies for those factors may aid in control. Anecdotally, control of other infections (PRRSV, parvovirus, others) or altering timing of administration of vaccines has been of benefit in reducing the severity of PCVAD. Studies in Denmark have shown that that PCVAD can be eradicated from an affected farm by total depopulation. The ubiquitous nature of PCV2 makes it unlikely that commercial farms can be maintained free of the virus by this technique but relief from clinical effects of PCVAD can likely be achieved. Until the epidemiology, transmission, and virulence factors relating to PCV2 and PCVAD are further understood, use of total depopulation as a control strategy is not often recommended.

Porcine circovirus Type 2 has been identified in the semen of acutely affected boars. However, transmission of the virus (or PCVAD) via this route has not been documented in a field setting.

Porcine circovirus Type 2 vaccines became commercially available in the US in summer of 2006. Field experiences have demonstrated that the vaccines are effective in reducing the severity and incidence rate of PCVAD on many farms. The timing of vaccination and population to be immunized (breeding herds, replacement stock, growing pigs) varies between farms depending on the characteristics of the affected populations and the physical operation of the farm.

Beak and Feather Diseases Viruses

Psittacine Beak and Feather Disease (PBFD) is a potentially deadly disease that affects parrots, cockatoos and lorikeets (psittacine birds). It is caused by the highly infectious Beak and Feather Disease (BFD) virus. It is not known to cause disease in humans.

PBFD is present in South Australia and all other states and territories in Australia. The first known outbreak of this disease was reported in wild Red-rumped parrots in the Adelaide Hills in 1888. It may have originated in Australia and is now widespread.

Symptoms

Symptoms vary greatly depending on the species and age of bird. The virus may affect the feathers, beak and claw and suppress the immune system.

PBFD should be considered in any parrot, cockatoo or lorikeet showing abnormal loss, colour or development of feathers. Cockatoos, Galahs and Little corellas may also develop abnormal beaks. The majority of affected birds will eventually die from secondary infections. PBFD can cause high mortality rates in young birds less than two years old.

Risk

PBFD does not present a major threat to the conservation of wild parrots, cockatoos and lorikeets unless there are only a few populations or limited numbers of birds, such as the critically endangered Orange-bellied parrot (Neophema chrysogaster) and the endangered Swift parrot (Lathamus discolour). PBFD has been recorded in wild birds of both species and its occurrence poses a risk to the survival of these species.

The endangered Glossy black cockatoo (Calyptorhynchus lathami halmaturinus) is known to be susceptible to PBFD, but the disease is not known in the small population on Kangaroo Island.

The disease is more prevalent in species that are widespread, such as the Sulphur-crested cockatoo, Little corella and Galah. It is quite common for a flock of these birds to have one or more members visibly affected by PBFD. The disease is not often found in cockatiels.

Spreading of PBFD

The virus remains viable for many years. It is spread from bird to bird in feather dust, droppings, in crop secretions when feeding chicks, and through successive use of the nest hollow or box. Birds can contract the virus at feeding, roosting and watering sites.

Some birds, such as the Rainbow lorikeet, are able to recover from the disease. They become a carrier of the virus and will excrete and spread the virus for the rest of their life.

People that have been in contact with an infected bird can spread the disease through their clothing and hair.

Deformed feathers and lost tail feathers
in an Eastern rosella with PBFD.

Prevention OF PBFD

There is no treatment available for PBFD and eradication of the disease is not feasible. The preferred management strategy is preventing and slowing down the spread.

Wherever possible, isolate diseased birds to prevent spread of the virus to other healthy birds. Diseased birds may need to be euthanised for the protection of healthy birds, particularly if birds are weak.

Do not release captive parrots, cockatoos and lorikeets into the wild (unless tested for PBFD and the test results show that the bird does not carry the BFD virus).

Good hygiene and husbandry of parrots, cockatoos and lorikeets in captivity is essential to protect them and wild birds from PBFD by:

- Placing food and water receptacles away from possible bird droppings.
- Quarantine and monitor any cockatoo, parrot or lorikeet for two months before allowing it to enter your aviary with other parrots, cockatoos and lorikeets, in particular when they are younger than two years. Preferably have the birds tested for PBFD.
- Clean surfaces of your aviary regularly with soaps and detergent and disinfect with 2% Virkon-S solution if it has held an infected bird.

- Shower and change clothing after handling a bird suspected of having PBFD and before getting in close proximity to other cockatoos, parrots and lorikeets.

The critically endangered Orange-bellied parrot is susceptible to PBFD, which may threaten the survival of this species, especially while the wild population is in such perilously low numbers.

Porcine Circovirus

A Circovirus is a small ssDNA virus which can inhabit and infect several hosts. Porcine circoviruses commonly infect swine. There are two main serotypes: Porcine circovirus 1 (PCV1) and Porcine circovirus 2 (PCV2). There is current research into another emerging serotype, Porcine circovirus 3 (PCV3), which is very similar to PCV2, but is not yet an epidemic in swine. PCV1 is not known to demonstrate any sign of disease, whereas PCV2 can cause an illness called Porcine circovirus Associated Disease (PCVAD), which causes Post Weaning Multisystemic Wasting Syndrome (PMWS). Porcine circovirus has one of the highest evolution rates of DNA viruses and also has the capability to co-infecting hosts along with other pathogens. This can lead to more severe outbreaks and complex syndromes involving reproductive failure, enteritis and pneumonia.

The Porcine circovirus has the smallest genome of any autonomously replicating virus. The small size of the genome presents many challenges when attempting to study the pathology of the virus in livestock. The circovirus can affect a variety of hosts (mostly livestock); moreover, Porcine circovirus infection can have detrimental effects on agricultural business as it can kill off an entire swine herd, and hence cut into profits. Infection results in a high fatality rate in swine populations, even though a vaccine has been created to protect against PMWS. The circovirus' small genome allows for fast evolutionary adaption, permitting the virus to quickly gain resistance to these vaccines. This is due to the fact that one or two mutations could have a huge effect on the virus, such as making a benign circovirus pathogenic.

Genome Structure

The Porcine circovirus has two major serotypes, type 1 and type 2 (PCV1 and PCV2) and one emerging serotype, type 3 (PCV3). PCV1 is known to be benign, while PCV2 is pathogenic. Porcine circovirus 2 (PCV2) is a small, non-enveloped virus with a circular single-stranded DNA genome which is 1.76 kb. Since the genome of circovirus is small it also allows for more recombination events with other similar viruses. For example, it has been shown historically through phylogenetic analysis that a nanovirus and a circovirus at some point recombined, as a circovirus has some RNA segments in its genome even though it is a ssDNA virus. The DNA sequence similarity between the two serotypes, PCV1 and PCV2, is about 76%.

Cell Structure

Porcine circovirus has a diameter of around 17 nm, thus making it the smallest animal virus studied with the capability of independent replication. Porcine circovirus is a non-enveloped virus assembled into a single capsid.

Metabolic Processes

Viruses do not have their own metabolism, but instead take advantage of their host cells to replicate, transcribe, and translate. The PCV genome enters the host cell nucleus by inserting into daughter nuclei at the end of mitosis. The two proteins Rep and Rep' are the replication initiator of Open Reading Frame (ORF) C1 transcript, which code for the capsid protein for porcine circovirus.

Ecology

Porcine circovirus 2 (PCV2) is highly infectious to swine. Studies determined that Porcine circovirus is a derivative of a former swine virus. The genome of the microbe and the cap (capsid) sequences of PCV2a and PCV2b viral genomes were aligned to determine any possible location of recombination in the sequence. Furthermore, circoviruses have been coevolving with many different vertebrate organisms for millions of years. In fact, the current evolutionary model of the Porcine circovirus is as follows: PCV2 is a virus that has long been associated with swine, and only recently has become infectious.

Pathology

Evidence shows that Porcine circovirus acts as an immunosuppressive agent which can open up the host to a secondary infection. It can lead to the development of Porcine circovirus-associated diseases (PCVAD) which are connected to many varying swine diseases. The most common disease that develops, within swine, is Post Weaning Multisystemic Wasting Syndrome (PMWS). PMWS is commonly found in swine 6-8 weeks with enlarged lymph nodes, jaundice, and extreme loss of weight. Some swine show symptoms of respiratory distress and interstitial pneumonia. PMWS has a high fatality rate with infected swine. To prevent this disease vaccines and prevention strategies for PCV2a were developed. Prevention strategies included monitoring the health of weaning swine, avoid having a high density of swine together, and control the movement of swine between batches. This however has caused the evolution of the virus to serotype PCV2b in swine populations, thus reducing the potential for prevention.

Porcine circovirus is not a zoonotic agent, and therefore there is little concern for possible infection to humans.

Epidemiology

Porcine circovirus is a global disease. In the United States and Italy, PCV3 was isolated and identified in 2015. Researchers are currently working to reduce the rate of infection of PCV3 with epidemiological analysis of the virus. Ultimately, epidemiologists hope the virus will not spread as quickly as PCV2 and will therefore avoid becoming an epidemic in swine populations around the globe. As of 2017 Italy was the only EU country with PCV3 which seemed to have similar genetic properties to the virus found in the US.

Currently, researchers are trying to characterize the complete genogroup of Porcine circovirus and its associated disease which may be involved in the epidemic of swine populations in many countries. This will help establish the preventive methods like vaccine production to resist the global outbreak of Porcine circovirus pathogens.

There is also current research in the epidemiology of an emerging serotype PCV3, and through the use of epidemiological analysis will keep the virus from becoming a new epidemic like PCV2.

Chicken Anemia Virus

Chicken anaemia virus infection (known generally in the industry as CAV) is an acute viral infection of chickens that is found worldwide. The disease is not known to affect any other bird species, although antibodies have been found in Japanese (coturnix) quail. Prior to confirmation that the disease was in fact caused by a virus it was known as Chicken Anaemia Agent or CAA.

CAV can infect chickens of all ages but disease is only seen in young chickens and is characterised by depression, anaemia, inappetence, haemorrhage and a sudden rise in mortality. CAV depresses the immune system and therefore leaves affected birds more susceptible to other infections and mortality can often be a result of secondary infections.

What Causes Chicken Anaemia Virus Infection?

CAV is a small DNA virus. In healthy chicks, susceptibility to disease declines rapidly with age and chicks are resistant to the clinical signs of the disease at 2 weeks of age. The virus can be spread both vertically (from parents to offspring) and horizontally (between birds within a flock), via the faecal-oral route.

Infected birds are viremic (shed virus) for up to 35 days. Infected roosters will shed the virus in their semen and hens will shed the virus into eggs during this viremic period.

Chicks infected through their parents can spread virus to other susceptible chicks with which they have contact, either directly or indirectly.

Recovered or immunised birds have neutralising antibodies that protect them from further infection. Chicks from immune breeder hens will be protected by maternal antibody until their own age resistance develops. Protection by maternal antibodies can however be overcome if the chick is affected by another severe immunosuppressive disease, such as infectious bursal disease, Marek's disease or reticuloendotheliosis.

Prevention and Treatment of Chicken Anaemia Virus Infection

There is no specific treatment. Secondary bacterial infections may be treated with antibiotics and minimised through good biosecurity practices, including hygiene and management. Vaccination of antibody-negative breeder flocks prior to the start of egg production is recommended. The control of other diseases that suppress the immune system is also important. At present, there is no vaccine available to prevent subclinical losses in broilers.

References

- Types-of-Parvovirus, health: news-medical.net, Retrieved 28 June, 2019
- Feline-panleukopenia, PetCare: avma.org, Retrieved 19 May, 2019
- Canine-parvovirus, PetCare: avma.org, Retrieved 04 July, 2019
- Goose-parvovirus-derzsys-disease, disease-guide: thepoultrysite.com, Retrieved 08 March, 2019
- Circovirus, FSVD-swine-index-diseases: vetmed.iastate.edu, Retrieved 15 April, 2019
- Porcine-Circovirus: microbewiki.kenyon.edu, Retrieved 23 January, 2019
- Chicken-anaemia-virus-infection-or-cav, health-disease-types-of-disease: poultryhub.org, Retrieved 05 February, 2019

Permissions

All chapters in this book are published with permission under the Creative Commons Attribution Share Alike License or equivalent. Every chapter published in this book has been scrutinized by our experts. Their significance has been extensively debated. The topics covered herein carry significant information for a comprehensive understanding. They may even be implemented as practical applications or may be referred to as a beginning point for further studies.

We would like to thank the editorial team for lending their expertise to make the book truly unique. They have played a crucial role in the development of this book. Without their invaluable contributions this book wouldn't have been possible. They have made vital efforts to compile up to date information on the varied aspects of this subject to make this book a valuable addition to the collection of many professionals and students.

This book was conceptualized with the vision of imparting up-to-date and integrated information in this field. To ensure the same, a matchless editorial board was set up. Every individual on the board went through rigorous rounds of assessment to prove their worth. After which they invested a large part of their time researching and compiling the most relevant data for our readers.

The editorial board has been involved in producing this book since its inception. They have spent rigorous hours researching and exploring the diverse topics which have resulted in the successful publishing of this book. They have passed on their knowledge of decades through this book. To expedite this challenging task, the publisher supported the team at every step. A small team of assistant editors was also appointed to further simplify the editing procedure and attain best results for the readers.

Apart from the editorial board, the designing team has also invested a significant amount of their time in understanding the subject and creating the most relevant covers. They scrutinized every image to scout for the most suitable representation of the subject and create an appropriate cover for the book.

The publishing team has been an ardent support to the editorial, designing and production team. Their endless efforts to recruit the best for this project, has resulted in the accomplishment of this book. They are a veteran in the field of academics and their pool of knowledge is as vast as their experience in printing. Their expertise and guidance has proved useful at every step. Their uncompromising quality standards have made this book an exceptional effort. Their encouragement from time to time has been an inspiration for everyone.

The publisher and the editorial board hope that this book will prove to be a valuable piece of knowledge for students, practitioners and scholars across the globe.

Index

www.ingramcontent.com/pod-product-compliance
Lightning Source LLC
LaVergne TN
LVHW080608210926
848485LV00003B/120